Lecture Notes in Physics

Volume 1032

The series Lecture Notes in Physics (LNP), founded in 1969, reports new developments in physics research and teaching - quickly and informally, but with a high quality and the explicit aim to summarize and communicate current knowledge in an accessible way. Books published in this series are conceived as bridging material between advanced graduate textbooks and the forefront of research and to serve three purposes:

- to be a compact and modern up-to-date source of reference on a well-defined topic;
- to serve as an accessible introduction to the field to postgraduate students and non-specialist researchers from related areas;
- to be a source of advanced teaching material for specialized seminars, courses and schools.

Both monographs and multi-author volumes will be considered for publication. Edited volumes should however consist of a very limited number of contributions only. Proceedings will not be considered for LNP.

Volumes published in LNP are disseminated both in print and in electronic formats, the electronic archive being available at springerlink.com. The series content is indexed, abstracted and referenced by many abstracting and information services, bibliographic networks, subscription agencies, library networks, and consortia.

Proposals should be sent to a member of the Editorial Board, or directly to the responsible editor at Springer:

Dr Lisa Scalone
lisa.scalone@springernature.com

Jonathan V. Selinger

Introduction to Topological Defects and Solitons

In Liquid Crystals, Magnets, and Related Materials

Springer

Jonathan V. Selinger
Department of Physics,
Advanced Materials and Liquid Crystal
Institute
Kent State University
Kent, OH, USA

ISSN 0075-8450 ISSN 1616-6361 (electronic)
Lecture Notes in Physics
ISBN 978-3-031-70199-3 ISBN 978-3-031-70200-6 (eBook)
https://doi.org/10.1007/978-3-031-70200-6

This Springer imprint is published by the registered company Springer Nature Switzerland AG
The registered company address is: Gewerbestrasse 11, 6330 Cham, Switzerland

Dedicated to my parents,
Anne and Carl Selinger,
in loving memory

Preface

In recent years, the study of topological defects and solitons has become an important topic of condensed-matter physics. Much of this progress has been in the fields of liquid crystals and magnetism, but it has also had impact on related areas, including biological physics.

In writing this book, my goal is to explain topological defects and solitons in a way that is accessible to a wide audience of students and researchers. I attempt to avoid difficult mathematics, as much as possible, and instead concentrate on the physical features of these structures.

To achieve that goal, I begin with a single physical model, the xy model for two-dimensional (2D) polar order. This model is easily visualized, it has a relatively simple mathematical description, and it has clear analogies with electrostatics. In Part I, the chapters cover many aspects of this model—first the topology of defects and solitons, then the free energy, and later dynamics and statistical mechanics. A further chapter shows how the model can be modified to describe polar order with variable magnitude, and uses this theory to discuss the relationship between topology and physics. Part I closes with a few more advanced points concerning defect orientation, topological charge density, and interaction of defects with curvature.

In Part II, the chapters generalize the concepts of topological defects and solitons to more complex physics systems. First, they discuss different types of order parameters—2D nematic order, 3D polar or nematic order, and crystalline order. Next, they go on to use different types of measuring surfaces to characterize defects and solitons, leading to hedgehogs (Bloch points), skyrmions, and hopfions. A final chapter considers the possibility of phases with regular arrays of defects or solitons, so that these topological structures actually occur in thermal equilibrium. All of these generalizations build on the fundamental principles established in Part I.

The book assumes that readers are familiar with multivariable calculus, i.e., partial derivatives, vector operators, and multiple integrals. It uses tensor notation occasionally, when necessary, but not very often. It would be helpful for readers to have some acquaintance with the statistical mechanics of phase transitions, perhaps from my previous book.[1] However, I try to summarize the necessary background information in this book, wherever it is needed.

If any readers would like to move on to more advanced study of topological defects and solitons, several excellent books and review articles are available. I especially recommend:

- N. D. Mermin, "The Topological Theory of Defects in Ordered Media," *Rev. Mod. Phys.* **51**, 591 (1979).
- M. Kleman, *Points, Lines and Walls: In Liquid Crystals, Magnetic Systems and Various Ordered Media* (Wiley, 1983).
- P. M. Chaikin and T. C. Lubensky, *Principles of Condensed Matter Physics* (Cambridge University Press, 1995).
- M. Kleman and O. D. Lavrentovich, *Soft Matter Physics: An Introduction* (Springer, 2003).
- G. P. Alexander, B. G. Chen, E. A. Matsumoto, and R. D. Kamien, "Colloquium: Disclination Loops, Point Defects, and All That in Nematic Liquid Crystals," *Rev. Mod. Phys.* **84**, 497 (2012).

Also, historical perspectives are provided by:

- T. J. Sluckin, "Some Reflections on Defects in Liquid Crystals: From Amerio to Zannoni and Beyond," *Liq. Cryst.* **45**, 1894 (2018).
- T. J. Sluckin, "Les Vertus Des Défauts: The Scientific Works of the Late Mr Maurice Kleman Analysed, Discussed and Placed in Historical Context, with Particular Stress on Dislocation, Disclination and Other Manner of Local Material Disbehaviour," *Liq. Cryst. Rev.* **11**, 74 (2023).

Other references are cited at appropriate points in the text.

I would certainly like to thank the people who have helped me to learn about this subject. I first studied topological defects and solitons with my PhD advisor, David Nelson. More recently, I have benefited greatly from discussions with several colleagues, especially Mark Bowick, Randall Kamien, Oleg Lavrentovich, and Ivan Smalyukh, and I have had a long-term collaboration with Robin Selinger. Needless to say, any errors are my own responsibility.

The original concept for this book was developed in an online program on "Symmetry, Thermodynamics and Topology in Active Matter" at the Kavli Institute for Theoretical Physics, University of California, Santa Barbara, supported in part by the National Science Foundation under Grant No. NSF PHY-1748958. The first version was presented as a lecture at the Center for Physics and Chemistry of Living Systems, Tel Aviv University, where I was visiting as a Distinguished Scholar of the Mortimer and Raymond Sackler Institute of Advanced Studies. I thank these

[1] J. V. Selinger, *Introduction to the Theory of Soft Matter: From Ideal Gases to Liquid Crystals* (Springer, 2016).

organizations for their sponsorship. I further thank my own institution, Kent State University, for its consistent support of my research and teaching over 19 years.

Kent, OH, USA
June 2024

Jonathan V. Selinger

Contents

Part I

Fundamental Principles

Introduction to Defects 1

Topology refers to aspects of geometric objects that are *counted*, rather than *measured*. For example, we might count that an object has zero or one or two holes, but we measure that its length is 2.79138 cm. The difference between counting and measuring is that counting gives an integer (with no units), while measuring produces a real number (with or without units). This difference is important if we make small changes to the object. As we continuously change the object, the measured length can change continuously to other real numbers. By contrast, the counted number of holes cannot change continuously. It must be locked at one integer unless it suddenly jumps to another integer. For that reason, a topological quantity like the number of holes is especially robust; it remains fixed even as the object changes.

In recent years, researchers have found that topology is useful in many areas of physics. Indeed, the application of topology to physics was recognized by the 2016 Nobel Prize in Physics to David Thouless, Duncan Haldane, and Michael Kosterlitz. One important example of this work is in the study of ordered states in condensed matter physics. Many condensed matter systems, including liquid crystals and magnets, have phase transitions from a disordered state (at high temperature) to an ordered state (at low temperature). The ordered state is characterized by an order parameter, which has both a magnitude and a direction. In general, the magnitude is determined by temperature, but the direction is arbitrary. In many cases, the direction is not uniform, but rather it varies across the sample, as a function of position. This nonuniform configuration can be characterized by topology. Indeed, the nonuniform configuration often has distinctive localized features, which can be described as topological defects or solitons. Those structures are the subject of this book.

Apart from topological defects and solitons in real space (position space), there are also other types of condensed matter with topological features in Fourier space (wave vector space). Those types of topological matter are not discussed in this book.

J. V. Selinger, *Introduction to Topological Defects and Solitons*, Lecture Notes in Physics 1032, https://doi.org/10.1007/978-3-031-70200-6_1

In the first six chapters, we consider a single physical example where the order parameter is a vector in two dimensions (2D). This example is commonly known as the xy model for 2D polar order, and it is the simplest possible model that produces topological defects and solitons. In spite of its simplicity, it demonstrates many fundamental physical principles, including the basic distinction between defects and solitons. After that, we consider generalizations to more complex systems: other types of order parameters (Chaps. 7–9), other types of measuring surfaces (Chaps. 10 and 11), and equilibrium phases with regular arrays of defects or solitons (Chap. 12). Those examples have many theoretical subtleties, but they still show the fundamental principles illustrated by the xy model.

1.1 Winding Number in 1D

Our standard example is the xy model, which represents a system that can develop order in any direction in a 2D plane. This type of order is called 2D polar order; it is a specific case of the general concept of *orientational order*. As an application of this model, we can consider a magnet, where the vector order parameter is the magnetic moment $\boldsymbol{M} = (M_x, M_y)$. In this example, some anisotropy forces $\boldsymbol{M}$ to lie in the (x, y) plane, perpendicular to the z axis. As an alternative application, we can consider a superconductor or superfluid, with a complex order parameter ψ. Here, the real and imaginary parts of ψ are the two dimensions of order. For now, we will use the magnetic terminology.

At high temperature, the system is in a disordered phase with $\boldsymbol{M} = 0$. At lower temperature, the system undergoes a transition into an ordered phase with $\boldsymbol{M} \neq 0$. The nonzero vector $\boldsymbol{M}$ has a magnitude and a direction. To emphasize these two aspects of $\boldsymbol{M}$, we write it as

$$\boldsymbol{M} = M\hat{\boldsymbol{n}}, \tag{1.1}$$

where $\hat{\boldsymbol{n}}$ is a unit vector in the 2D plane. As a unit vector, it can be written as

$$\hat{\boldsymbol{n}} = (\cos\theta, \sin\theta) \tag{1.2}$$

for some angle θ. This angle is defined modulo 2π; i.e. the angles θ and $\theta + 2\pi$ represent the same physical state.

With this separation into magnitude and direction, let us first consider the magnitude M. The system tends to minimize the free energy $F = E - TS$, which is a combination of energy E, temperature T, and entropy S. At any specific T, this minimization gives a certain equilibrium magnitude M. It would cost a large amount of free energy to change M away from this equilibrium magnitude. Hence, for now, we will just assume that the magnitude M is constant. We will not think about it again until Chap. 5.

Unlike the magnitude, the direction $\hat{\boldsymbol{n}}$ is *not* determined by minimization of the free energy. The free energy has rotational symmetry, which means that it does not depend on the direction of the magnetic order. Any direction in the (x, y) plane gives the same free energy. Hence, the direction is selected by *spontaneous symmetry breaking*—the system spontaneously develops magnetic order in some arbitrary direction. Indeed, at different positions, the system might develop magnetic order in different directions. Thus, $\hat{\boldsymbol{n}}$ generally depends on position $\boldsymbol{r}$, and we must treat it as a function $\hat{\boldsymbol{n}}(\boldsymbol{r})$.

Although the free energy does not depend on the direction $\hat{\boldsymbol{n}}$, it might depend on derivatives of $\hat{\boldsymbol{n}}$ with respect to position. This dependence will be discussed in Chap. 3. For now, we can just note that, if $\hat{\boldsymbol{n}}$ has a sharp jump from one position to another, that jump would cost a large amount of free energy. Hence, we can assume that $\hat{\boldsymbol{n}}(\boldsymbol{r})$ is a smooth, continuous function of position.

We would now like to consider the possible global configurations of $\hat{\boldsymbol{n}}(\boldsymbol{r})$ as a function of position. As a first step, suppose we have a *finite, one-dimensional (1D)* system. Here, *one-dimensional* means that the position $\boldsymbol{r}$ has only one component, which is the coordinate x. The unit vector $\hat{\boldsymbol{n}}(x) = (\cos\theta(x), \sin\theta(x))$ still has two components. *Finite* means that the system has boundaries at $x_{\min}$ and $x_{\max}$. At the boundaries, it may have boundary conditions on $\hat{\boldsymbol{n}}(\mathrm{x})$. To be specific, suppose that $\hat{\boldsymbol{n}}(x_{\min}) = \hat{\boldsymbol{n}}(x_{\max}) = \hat{\boldsymbol{y}}$. Hence, $\hat{\boldsymbol{n}}(x)$ must satisfy three requirements:

1. It is a unit vector, with $|\hat{\boldsymbol{n}}(x)| = 1$ for all x.
2. It is a smooth, continuous function of x.
3. It goes to $\hat{\boldsymbol{y}}$ at the two boundaries, $x \to x_{\min}$ and $x \to x_{\max}$.

Figure 1.1 shows several examples of $\hat{\boldsymbol{n}}(x)$ configurations that satisfy all three requirements. In this figure, the dark blue arrows represent the unit vector $\hat{\boldsymbol{n}}(x)$ on a regularly spaced lattice of positions from $x_{\min}$ to $x_{\max}$. It is important to remember that $\hat{\boldsymbol{n}}(x)$ is defined everywhere along the 1D line—not just on the lattice positions, but also in the gaps between them. As a reminder, in this figure, we have drawn light blue arrows to fill in the gaps. In later figures, we will omit the light blue arrows because they make the figures too cluttered, but readers should always remember to mentally fill in the gaps between the lattice points.

To characterize the $\hat{\boldsymbol{n}}(x)$ configurations, we construct the *winding number*.[1] The winding number q is defined as the number of times that $\hat{\boldsymbol{n}}(x)$ rotates through a full circle, in a counter-clockwise direction, as we travel from $x_{\min}$ to $x_{\max}$. It can be determined by adding up all of the small increments of the angle θ as we move along the line. Mathematically, it can be expressed as the integral

$$q = \frac{1}{2\pi} \int_{x_{\min}}^{x_{\max}} d\theta = \frac{1}{2\pi} \int_{x_{\min}}^{x_{\max}} \left(\frac{d\theta}{dx} \right) dx. \tag{1.3}$$

[1] English language note: In this context, the word *wind* is pronounced with a long *i*, so that it rhymes with *find*, *kind*, and *mind*. It should not be confused with the word *wind* with a short *i*, which means moving currents of air.

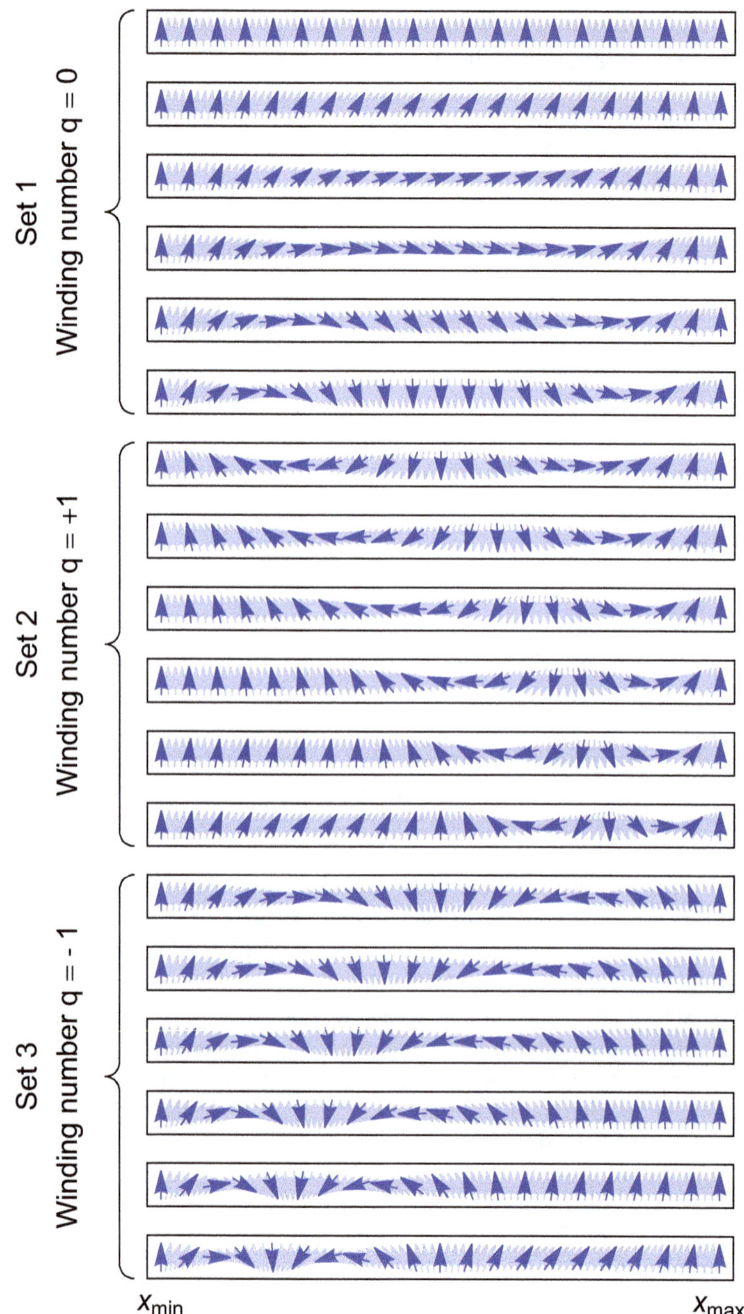

Fig. 1.1 Several examples of $\hat{\boldsymbol{n}}(x)$ configurations in a finite 1D system from $x_{\min}$ to $x_{\max}$. The winding number is $q = 0$ for the configurations in set 1, $q = +1$ for the configurations in set 2, and $q = -1$ for the configurations in set 3

If we want to work with the unit vector $\hat{\boldsymbol{n}}(x)$ instead of the angle $\theta(x)$, we can just use the vector equation

$$\hat{\boldsymbol{n}} \times \frac{d\hat{\boldsymbol{n}}}{dx} = (\cos\theta, \sin\theta) \times (-\sin\theta, \cos\theta)\frac{d\theta}{dx} = \hat{z}\left(\frac{d\theta}{dx}\right), \tag{1.4}$$

so that the winding number becomes

$$q = \frac{1}{2\pi}\int_{x_{\min}}^{x_{\max}} \hat{z} \cdot \left(\hat{\boldsymbol{n}} \times \frac{d\hat{\boldsymbol{n}}}{dx}\right) dx. \tag{1.5}$$

We apply this concept to the configurations in Fig. 1.1:

- Set 1: In all of the configurations, $\hat{\boldsymbol{n}}(x)$ rotates by equal amounts counter-clockwise (with $d\theta > 0$) and counter-clockwise ($d\theta < 0$). All of these changes in θ add up to zero, so that $\theta(x_{\max}) = \theta(x_{\min})$, and hence $\hat{\boldsymbol{n}}(x_{\max}) = \hat{\boldsymbol{n}}(x_{\min})$. Hence, the winding number is $q = 0$.
- Set 2: In all configurations, $\hat{\boldsymbol{n}}(x)$ rotates more counter-clockwise than clockwise. Adding up these increments gives $\theta(x_{\max}) = \theta(x_{\min}) + 2\pi$, and this increase of 2π gives $\hat{\boldsymbol{n}}(x_{\max}) = \hat{\boldsymbol{n}}(x_{\min})$. These configurations have $q = +1$.
- Set 3: In all of these cases, $\hat{\boldsymbol{n}}(x)$ rotates more clockwise than counter-clockwise. The integral of these changes is $\theta(x_{\max}) = \theta(x_{\min}) - 2\pi$, which again gives $\hat{\boldsymbol{n}}(x_{\max}) = \hat{\boldsymbol{n}}(x_{\min})$. All of these configurations have $q = -1$.

In general, the difference $\theta(x_{\max}) - \theta(x_{\min})$ must be an integer multiple of 2π, so that $\hat{\boldsymbol{n}}(x_{\max}) = \hat{\boldsymbol{n}}(x_{\min})$, in order for the system to satisfy the boundary conditions. Hence, the winding number must be an integer—positive, negative, or zero.

Now suppose that we make some small, gradual changes to the configuration of $\hat{\boldsymbol{n}}(x)$, consistent with the three requirements (a)–(c) above. Because the winding number depends smoothly on $\hat{\boldsymbol{n}}(x)$, it can only change in a gradual way. However, it is impossible to make a gradual change to a quantity that must be an integer. It can only jump discontinuously to another integer. Hence, the winding number must remain constant, at its original integer value. This property is the essence of a topological quantity. We can say that the winding number is *topologically protected*, meaning that it remains constant under gradual changes of the $\hat{\boldsymbol{n}}(x)$ configuration.

Based on this topological property of the winding number, we can see that it is possible to transform any configuration in Set 1 to any other configuration in Set 1, using gradual changes to the $\hat{\boldsymbol{n}}(x)$ configuration, consistent with the three requirements (a)–(c). Likewise, it is possible to transform any configuration in Set 2 to any other configuration in Set 2, or any configuration in Set 3 to any other configuration in Set 3. However, it is impossible to transform a configuration in one set to a configuration in another set using gradual changes to the $\hat{\boldsymbol{n}}(x)$ configuration, because such a transformation would require a discontinuous jump in the winding number.

1.2 Singular Point in 2D

We now generalize the concepts from the previous section to a *two-dimensional (2D)* system. *Two-dimensional* means that the position $\boldsymbol{r} = (x, y)$ has two components. Of course, the unit vector $\hat{\boldsymbol{n}}(\boldsymbol{r}) = (\cos\theta(\boldsymbol{r}), \sin\theta(\boldsymbol{r}))$ also has two components. Suppose that the 2D system is finite in the x direction, running from x_{min} to x_{max}, and it has the boundary conditions $\hat{\boldsymbol{n}} = \hat{\boldsymbol{y}}$ at $x = x_{\text{min}}$ and $x = x_{\text{max}}$, for all y. For this discussion, we do not care about boundaries in the y direction. One example of an $\hat{\boldsymbol{n}}(\boldsymbol{r})$ configuration satisfying the boundary conditions is shown in Fig. 1.2.

In 2D, the winding number must be defined along a path running across the system, from x_{min} to x_{max}. The path may be straight and horizontal (like the top green line in Fig. 1.2), or diagonal (middle green line), or curved in a complicated way (bottom green line). Along this path, we add up all of the small increments of the angle θ, and we obtain the line integral

$$q = \frac{1}{2\pi}\int_{\text{path}} d\theta = \frac{1}{2\pi}\int_{\text{path}} (\nabla\theta)\cdot d\boldsymbol{l}. \tag{1.6}$$

In terms of $\hat{\boldsymbol{n}}(\boldsymbol{r})$ rather than $\theta(\boldsymbol{r})$, the line integral becomes

$$q = \frac{1}{2\pi}\int_{\text{path}} (\epsilon_{3ij} n_i \partial_k n_j) dl_k. \tag{1.7}$$

This expression uses tensor notation, with an implicit summation over indices i, j, k, and ϵ_{3ij} is the Levi-Civita symbol.

Let us consider how the winding number might depend on the choice of path. The $\hat{\boldsymbol{n}}(\boldsymbol{r})$ configuration along any specific path is analogous to the configuration in a 1D

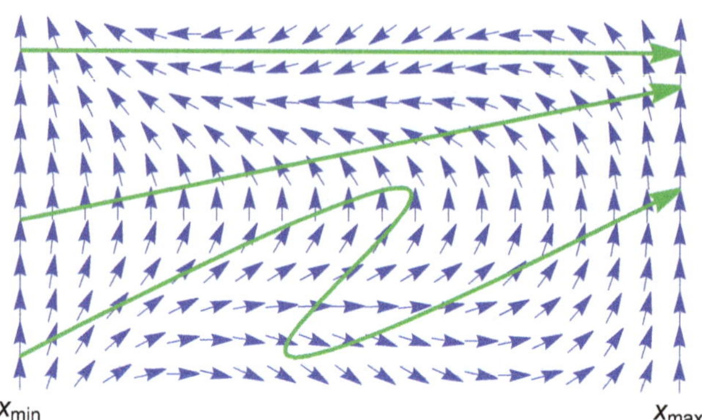

Fig. 1.2 Example of a configuration for $\hat{\boldsymbol{n}}(\boldsymbol{r})$ in a 2D system, with no singularities. The green lines show three examples of paths across the system, from x_{min} to x_{max}. The winding number along any of these paths is $q = 0$

system, as discussed in the previous section. It must begin at $\hat{\boldsymbol{n}}(x_{\min}) = \hat{\boldsymbol{y}}$ and end at $\hat{\boldsymbol{n}}(x_{\max}) = \hat{\boldsymbol{y}}$. To satisfy these boundary conditions, the difference $\theta(x_{\max}) - \theta(x_{\min})$ must be an integer multiple of 2π, and the winding number must be an integer. Suppose we make small, gradual changes to the path. *If the $\hat{\boldsymbol{n}}(\boldsymbol{r})$ configuration is well-defined throughout the whole range of paths,* then we are effectively making gradual changes to the configuration in the 1D system. Because the winding number depends smoothly on $\hat{\boldsymbol{n}}(\boldsymbol{r})$, it must also change in a gradual way. Because it cannot change gradually from one integer to another, it must remain at its original integer value, independent of the path. Indeed, Fig. 1.2 provides an example of this behavior. Here, the winding number is $q = 0$ for all of the green paths that are shown, and for any other path from $x_{\min}$ to $x_{\max}$.

A very different example of an $\hat{\boldsymbol{n}}(\boldsymbol{r})$ configuration is shown in Fig. 1.3. This example has a *singular point*, or *singularity*, indicated by the red dot. The unit vector points outward in all directions from the singularity. Hence, $\hat{\boldsymbol{n}}(\boldsymbol{r})$ is undefined right at the singularity, but it is well-defined everywhere else in the system. The presence of a singularity has important consequences for the winding number. If we integrate along any path *below* the singularity (like the lower three green lines in Fig. 1.3a), the winding number is $q = +1$. By contrast, if we integrate along any path *above* the singularity (upper three green lines in Fig. 1.3a), the winding number is $q = 0$. If the path passes directly *through* the singularity, the winding number is undefined, because $\hat{\boldsymbol{n}}(\boldsymbol{r})$ and its derivatives are undefined there. Hence, in this example, the winding number depends on the path—it changes from +1 to undefined to 0 as we shift the path upward. This path-dependence does not violate the rule in the previous paragraph, as the $\hat{\boldsymbol{n}}(\boldsymbol{r})$ configuration is not well-defined throughout the whole range of paths.

Let us consider the composite path shown in Fig. 1.3b. This path begins at $x_{\min}$, moves straight toward the singularity, takes a half-loop *below* the singularity, then moves straight to $x_{\max}$. It then returns back toward the singularity, takes a half-loop *above* the singularity, then moves straight back to $x_{\min}$. For the first half of this path, the winding number is +1, because it passes below the singularity. For the second half (moving backward from $x_{\max}$ to $x_{\min}$), the line integral is the negative of the forward line integral, i.e. the negative of 0, and hence the winding number is 0. Hence, the total winding number for the composite path from $x_{\min}$ to $x_{\max}$ and back to $x_{\min}$ is $q = +1$.

Now we can simplify the path. On the left-hand side of the figure, between $x_{\min}$ and the half-loops, our path goes both forward and backward. These two contributions to the line integral cancel each other. Likewise, on the right-hand side of the figure, between the half-loops and $x_{\max}$, our path again goes both forward and backward, and those two contributions cancel each other. After removing those parts of the path, what remains is a *full measuring loop*, shown in Fig. 1.3c, moving in the *counter-clockwise* direction. The winding number around that full loop is $q = +1$.

For this argument, it does not matter whether the full loop is large or small. Also, it does not matter whether the full loop is a perfect circle or any other shape. This argument shows that the winding number is +1 around *any* loop that encloses the

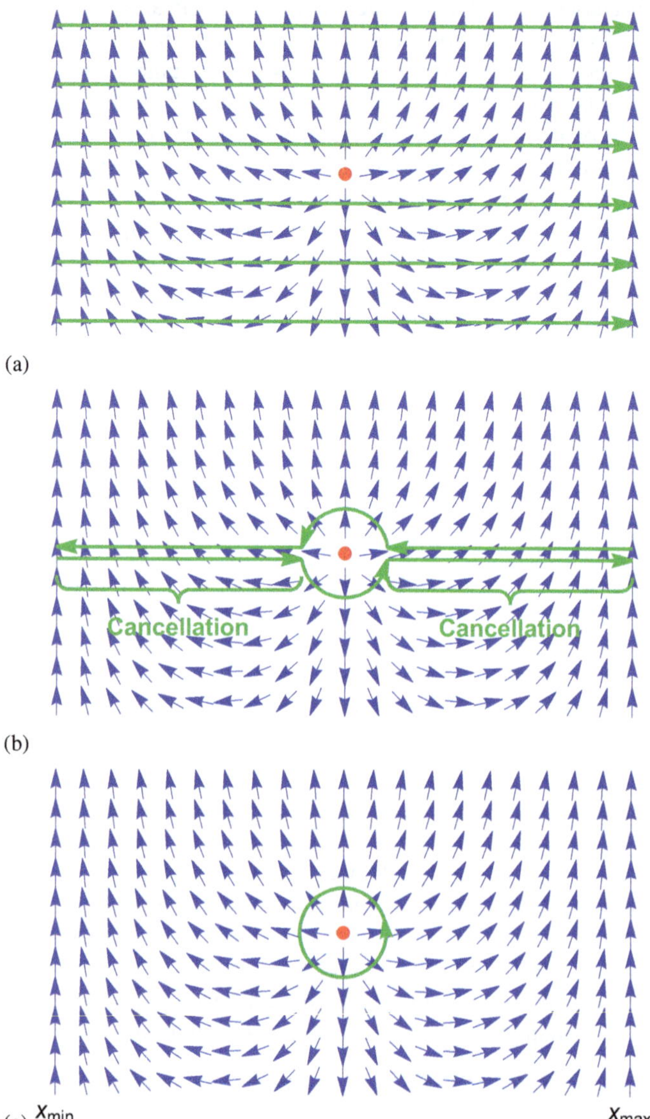

Fig. 1.3 Example of a configuration for $\hat{\boldsymbol{n}}(\boldsymbol{r})$ in a 2D system, with a singular point indicated by the red dot. (**a**) Several integration paths across the system, above and below the singular point. (**b**) Integration path that begins at x_{min}, makes a half-loop *below* the singular point, goes on to x_{max}, and then returns with a half-loop *above* the singular point. (**c**) Integration loop around the singular point

singularity. By comparison, through an analogous argument, the winding number is 0 around any loop that does not enclose the singularity.

At this point, we make an intellectual leap: Instead of saying that the winding number is a property of the path, we can say that the winding number is a property of the singularity. In other words, we will characterize the singularity by its effect on the $\hat{\boldsymbol{n}}(\boldsymbol{r})$ configuration, defined by the line integral on a loop around the singularity. In this way of thinking, the singularity is a *topological defect*, which has a *topological charge* given by

$$q_{\text{enclosed}} = \frac{1}{2\pi}\oint d\theta = \frac{1}{2\pi}\oint (\nabla\theta)\cdot d\boldsymbol{l} = \frac{1}{2\pi}\oint (\epsilon_{3ij} n_i \partial_k n_j) dl_k. \quad (1.8)$$

Here, the line integral is taken around any loop, in the counter-clockwise direction, and q_{enclosed} is the total topological charge enclosed by the loop. This equation is similar to Gauss's law in electrostatics, which relates the electric charge to an integral over a surface enclosing the charge. In that sense, the topological charge is analogous to an electric charge.

We have now defined topological defects by a loop construction inside a system, with no reference to the boundaries. Hence, we can discard the boundary conditions at $x_{\min}$ and $x_{\max}$, and just think about topological defects in an arbitrary geometry, or even in an infinite system. Figure 1.4 shows several examples of topological defects in the bulk, away from any boundary. The examples in Fig. 1.4a–d all have topological charge $q = +1$. Note that the unit vector $\hat{\boldsymbol{n}}(\boldsymbol{r})$ may point radially outward or inward, tangentially counter-clockwise or clockwise, or in any intermediate direction. In all of these cases, $\hat{\boldsymbol{n}}(\boldsymbol{r})$ rotates through a full circle *counter-clockwise* as we move around the loop *counter-clockwise*.[2]

By comparison, the example in Fig. 1.4e has topological charge $q = -1$. In this case, the unit vector $\hat{\boldsymbol{n}}(\boldsymbol{r})$ points radially outward along one axis, and radially inward along the orthogonal axis. It rotates through a full circle *clockwise* as we move around the loop *counter-clockwise*. The example in Fig. 1.4f has topological charge $q = +2$. It rotates through two full circles counter-clockwise as we move around the loop counter-clockwise. In general, there can be topological defects with a topological charge of any *positive or negative integer*.

For the xy model discussed here, topological charges add together like electric charges.[3] As an example, Fig. 1.5a shows a configuration with two topological defects. By integrating around the small loops, we see that each individual defect has a topological charge of +1. By integrating around the large loop, we see that the combination has a total topological charge of +2. Similarly, Fig. 1.5b has two

[2] Based on the analogy with Gauss's law, some readers might have expected the outward and inward structures in Fig. 1.4a and b to have opposite topological charges. However, that expectation is not correct, because the unit vector $\hat{\boldsymbol{n}}(\boldsymbol{r})$ is not analogous to electric field. In Sect. 6.2.1, we will discuss a different vector that is analogous to electric field.

[3] As a warning to readers, this simple addition rule works for topological defects in the xy model, but it will not work for all of the more complex models discussed in Part II of this book.

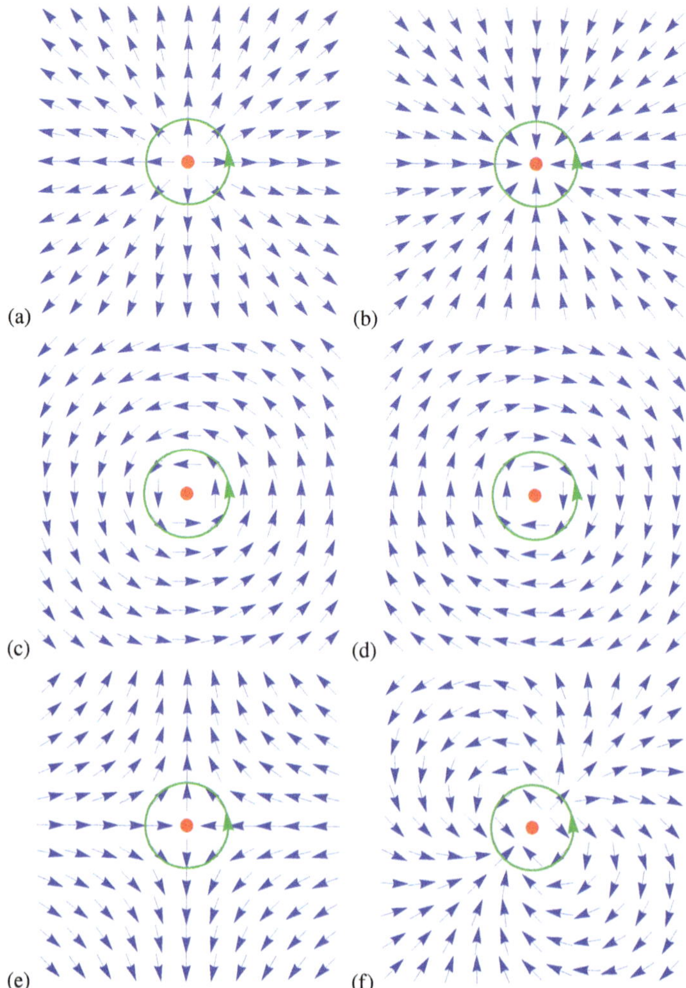

Fig. 1.4 Examples of topological defects in the bulk, away from any boundary. Examples (**a**)–(**d**) have topological charge $q = +1$, example (**e**) has topological charge $q = -1$, and example (**f**) has topological charge $q = +2$

defects of topological charge $+1$ and -1, respectively. By integrating around the large loop, we see that the combination has a total topological charge of 0. If these two defects move together, they would annihilate each other like a particle and an antiparticle, to leave an $\hat{\boldsymbol{n}}(\boldsymbol{r})$ configuration with no defects.

Different subfields of physics have different terminology for these topological defects. They may be called *disclinations* or *vortices*. Sometimes defects of topological charge $+1$ are called *vortices*, and defects of topological charge -1 are called *antivortices*. Sometimes defects with the tangential structure in Fig. 1.4c

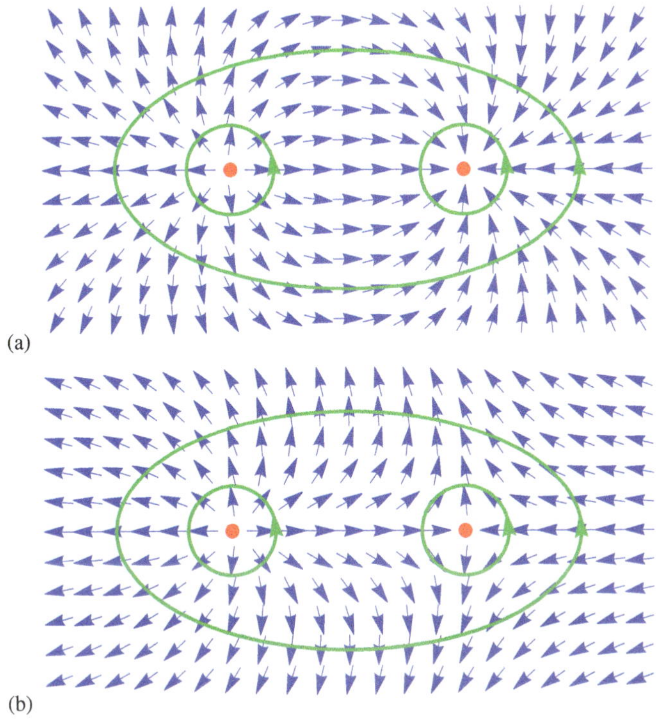

Fig. 1.5 Examples of $\hat{\boldsymbol{n}}(\boldsymbol{r})$ configurations with two topological defects. (**a**) Two topological charges of $+1$ each, which combine to make a total topological charge of $+2$. (**b**) Two topological charges of $+1$ and -1, which combine to make a total topological charge of 0

and d are called *vortices*, and defects with the radial structure in Fig. 1.4a and b are called *asters*.

1.3 Defect Line in 3D

Let us now move on to a *three-dimensional (3D)* system. *Three-dimensional* means that the position $\boldsymbol{r} = (x, y, z)$ has three components. The unit vector $\hat{\boldsymbol{n}}(\boldsymbol{r}) = (\cos\theta(\boldsymbol{r}), \sin\theta(\boldsymbol{r}))$ still has two components. (We will consider a 3D unit vector later, in Chap. 8.)

The concept of topological defects can be extended from 2D to 3D. As an example, consider the $\hat{\boldsymbol{n}}(\boldsymbol{r})$ configuration shown in Fig. 1.6a. Here, the unit vector $\hat{\boldsymbol{n}}(\boldsymbol{r})$ is well-defined everywhere in the system except along a *singular line*, indicated by the red line in the figure. To characterize the singularity, we can draw a measuring loop that passes around the singular line, moving in the counter-clockwise direction—for example, we could use any of the green loops in the figure. By integrating around any of those loops, we find the winding number $q = +1$. By

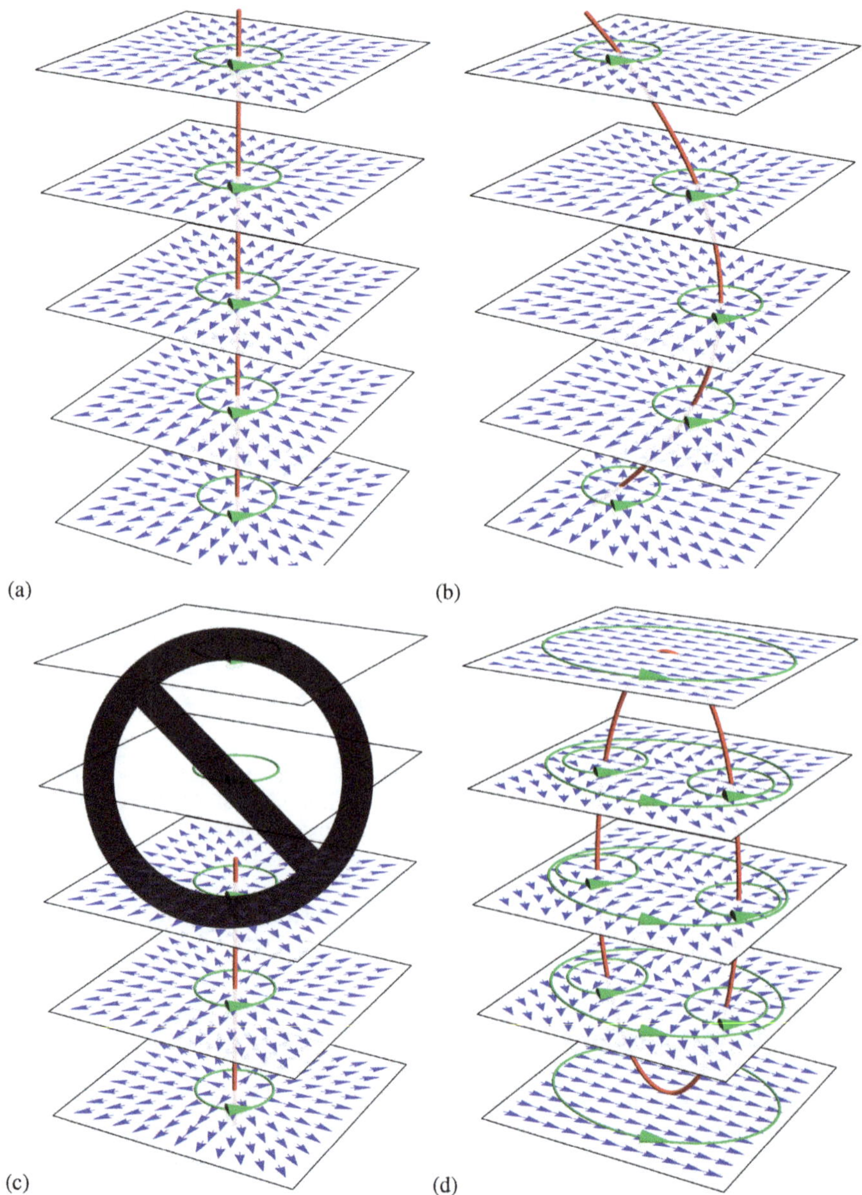

Fig. 1.6 Configurations of the 2D unit vector $\hat{\boldsymbol{n}}(\boldsymbol{r})$ in a 3D system. (**a**) Straight defect line with topological charge $+1$. (**b**) Curved defect line with topological charge $+1$. (**c**) *Impossible* structure with defect line ending inside the system. (**d**) Defect loop

comparison, for any loop that does *not* pass around the singular line, the winding number is $q = 0$. Hence, this configuration is a *defect line* with topological charge $q = +1$. Similarly, there can be defect lines with a topological charge of any positive or negative integer.

A defect line does not need to be straight. It might be curved, as illustrated in Fig. 1.6b. Once again, the winding number is $q = +1$ for any loop that goes around the singular line in the counter-clockwise direction, such as the green loops in the figure. It is $q = 0$ for any loop that does not go around the singular line. This configuration is a curved defect line, again with topological charge $q = +1$.

For this argument, it does not matter whether the measuring loop is large or small, or whether it is a circle or any other shape. Furthermore, it does not matter whether the loop is in the horizontal xy plane, or tilted out of the plane. We can make any gradual changes to the loop, and we will get the same winding number, as long as the loop does not cross the singular line where $\hat{\boldsymbol{n}}(\boldsymbol{r})$ is undefined. Hence, the topological charge q is a property of the defect line, not a property of the measuring loop.

One important requirement for defect lines is: *A defect line cannot begin or end anywhere inside a system.* To understand the reason for this requirement, suppose that it were not true. In that case, we could have structure as in Fig. 1.6c, with a defect line below some height z_0, and no defect line above z_0. If we draw a measuring loop around the defect line below z_0, we would find $q = +1$. If we gradually move the measuring loop upward, the winding number should not jump discontinuously anywhere, so it must remain $q = +1$ even above z_0. This result contradicts our assumption of no defect line above z_0, and we must conclude that the requirement is true.

Because a defect line cannot begin or end anywhere inside a system, it might go all the way to the boundary of the system. Alternatively, it might curve into a closed loop, as shown in Fig. 1.6d. In this example, the defect line does not have any beginning or end. If we draw any measuring loop that passes around the local defect line, we find a nonzero winding number: $q = +1$ for the small green loops on the right, $q = -1$ for the small green loops on the left. By comparison, if we draw any measuring loop that passes around the whole red defect loop structure, such as the large green loops, then we find $q = 0$.

Readers should not confuse the defect loop (shown in red) with the measuring loops (shown in green). The red defect loop is a physical object; it is the whole set of singular points where $\hat{\boldsymbol{n}}(\boldsymbol{r})$ is undefined. By comparison, the green measuring loops are just artificial constructions, which we can draw anywhere, in order to determine the winding number.

One way to interpret the defect loop of Fig. 1.6d, which might be interesting for some readers, is as *world lines* of a particle and an antiparticle. In special relativity, a world line represents the path of a particle in spacetime. Here, we can imagine that the z coordinate represents time, while the x and y coordinates are space. At early time (low z), the system has no defects. Moving forward in time (upward in z), we reach the bottom of the loop. Here, a particle ($+1$ defect) and antiparticle (-1 defect) spontaneously form and move apart. After some time (higher in z), they eventually come together and annihilate each other, and the system again has no defects.

1.4 Topological Notation

In this book, we generally do not use the mathematical terminology of topology. Instead, we try to provide non-mathematical explanations of topological concepts. However, readers should be familiar with one type of mathematical notation. The notation for the type of defects discussed in this chapter is:

$$\pi_1(S^1) = \mathbb{Z}. \tag{1.9}$$

Here, we explain the symbols in that expression.

The symbol S^1 describes the type of order parameter, which is the xy model. The letter S means *sphere*, and the superscript 1 means one-dimensional. A one-dimensional sphere is a circle.[4] This circle represents the set of all possible unit vectors $\hat{\boldsymbol{n}} = (\cos\theta, \sin\theta)$. Physically, it means all the possible local ground states of the order parameter. For example, if we are working with a 2D magnetic system, then it is all the possible orientations of the local magnetic moment. In Chaps. 7, 8, and 9, we will generalize to other types of order parameters beyond the xy model.

The symbol π_1 refers to the *first homotopy group*. It describes a classification of defects based on one-dimensional measuring surfaces, which are the green paths and green loops shown in the figures in this chapter. In this classification, we consider the winding number, which describes what happens to the order parameter as we move along a path or around a loop. In Chaps. 10 and 11, we will generalize to higher-dimensional measuring surfaces.

The symbol $\mathbb{Z}$ represents the set of all integers (positive, negative, and zero). Hence, Eq. (1.9) tells us: If we consider the xy model, and we examine what happens to $\hat{\boldsymbol{n}}(\boldsymbol{r})$ configurations along 1D paths and loops, the configurations can be classified by a single integer, which is the winding number. Any configuration can be gradually transformed into another configuration with the same winding number, but not into another configuration with a different winding number.

Using this mathematical notation, we can make one simple comment about the dimension of defects. In this chapter, we have already seen that the dimension of defects is related to the dimension of space. Section 1.2 showed that 2D space has point defects, which are zero-dimensional. Section 1.3 showed that 3D space has line defects, which are one-dimensional.

In general, suppose the dimension of space is d_{space}, and the dimension of the measuring surface is d_{measure}, and we want to determine the dimension of defects d_{defect}. The measuring surface encloses a region of dimension $d_{\text{measure}} + 1$. The

[4] Mathematically, a *circle* is a curved line, which is one-dimensional. It should not be confused with the area enclosed by a circle, called a *disk*, which is two-dimensional.

defect extends *across* that enclosed region, so it fills up all the dimensions of space *except* the dimensions of that enclosed region. Hence, the dimension of defects becomes

$$d_{\text{defect}} = d_{\text{space}} - d_{\text{measure}} - 1. \tag{1.10}$$

In later chapters, we will see this relationship in other contexts.

Introduction to Solitons

2

In the previous chapter, we presented topological defects, which are configurations of the unit vector $\hat{\boldsymbol{n}}(\boldsymbol{r})$ characterized by a winding number. In this chapter, we introduce another type of $\hat{\boldsymbol{n}}(\boldsymbol{r})$ configuration, called a topological soliton. As we will see, topological solitons are *also* characterized by a winding number, but they are *different* from topological defects.

2.1 Soliton in 1D

For the basic concept of a topological soliton, let us return to a 1D system, as in Sect. 1.1. Suppose the winding number is $q = 1$, meaning that $\hat{\boldsymbol{n}}(x)$ rotates through one full circle, in a counter-clockwise direction, as we travel from x_{min} to x_{max}. For all of the examples in Fig. 1.1, the rotation is spread out across the entire system. We now consider what happens if the rotation becomes concentrated in a narrow region of width ξ, as illustrated in Fig. 2.1. The width ξ does not need to be microscopic or molecular in scale. The important point is just that it is small compared with other lengths of interest—especially compared with the system size.

Readers might wonder: *Why* would the rotation become concentrated in a narrow region? We will postpone that question until Sect. 3.3. For now, let us just assume that the rotation has some natural length scale ξ, which is small.

The concentrated rotation of $\hat{\boldsymbol{n}}(x)$ is called a topological soliton. From this example, we can see several properties of solitons. First, a soliton is not a mathematical point, but it is a fairly narrow, localized region, which is highlighted in yellow. The soliton may be located anywhere in the interior of the system. Inside the soliton, the unit vector $\hat{\boldsymbol{n}}(x)$ changes rapidly as a function of position, rotating by approximately 2π over a distance of approximately ξ. Outside the soliton, $\hat{\boldsymbol{n}}(x)$ is approximately constant, at the background orientation of $\hat{\boldsymbol{x}}$.

Because the rotation of $\hat{\boldsymbol{n}}(x)$ is concentrated inside the soliton, the integral for the winding number in Eq. (1.5) is also concentrated inside the soliton. The background

J. V. Selinger, *Introduction to Topological Defects and Solitons*, Lecture Notes in Physics 1032, https://doi.org/10.1007/978-3-031-70200-6_2

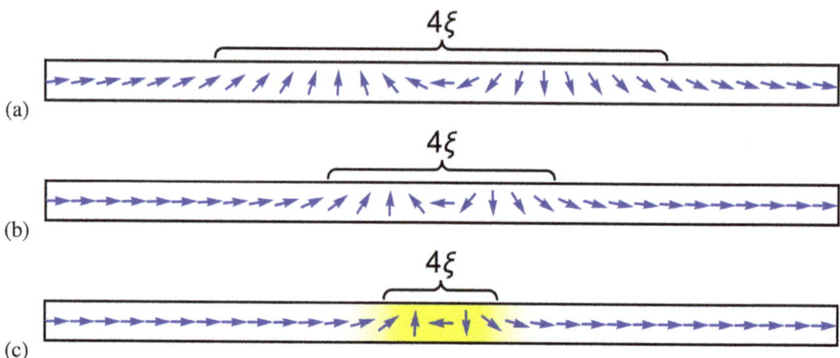

Fig. 2.1 Examples of $\hat{\boldsymbol{n}}(x)$ configurations in a 1D system with winding number $q = +1$. In example (**a**), the rotation is spread out over the entire width of the system. In example (**b**), it becomes more concentrated. In example (**c**), the rotation is highly concentrated in a characteristic length ξ, which is much less than the width of the system. The concentrated rotation is called a soliton, and is highlighted in yellow

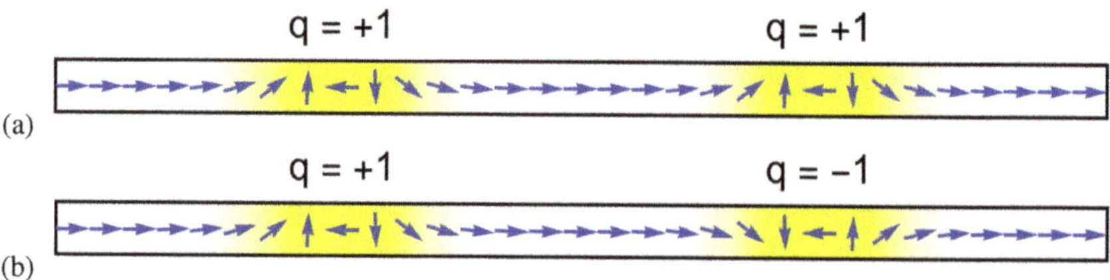

Fig. 2.2 Examples of $\hat{\boldsymbol{n}}(x)$ configurations with two solitons. (**a**) Two solitons, each with $q = +1$. (**b**) Solitons with $q = +1$ and $q = -1$, respectively

region outside the soliton contributes almost nothing to this integral. Hence, as an excellent approximation, we can say that the winding number is associated with the soliton, rather than with the entire path from x_{min} to x_{max}. In other words, the soliton is a localized variation of $\hat{\boldsymbol{n}}(x)$ with a winding number.

The winding number associated with a soliton may be positive or negative, depending on whether $\hat{\boldsymbol{n}}(x)$ rotates counter-clockwise or clockwise as x increases. The overall sign depends on the convention of which direction is labeled as positive x, but we can unambiguously say whether two solitons have the same or opposite sign. Figure 2.2a shows an example with two solitons, each with winding number $+1$. The combined system of both solitons has a total winding number of $+2$. By comparison, Fig. 2.2b shows one soliton of winding number $+1$ and another soliton of winding number -1. Here, the combined system has a total winding number of 0. Those two solitons could potentially move together and annihilate each other, leaving behind a system with the background orientation $\hat{\boldsymbol{n}}(x) = \hat{\boldsymbol{x}}$.

Another term for a soliton of winding number ± 1 is a 2π-wall, because $\hat{\boldsymbol{n}}(x)$ rotates through an angle of 2π across it. The $+1$ and -1 solitons can be regarded as a wall and an antiwall, like a particle and an antiparticle, because they can annihilate each other.

In principle, the winding number of a soliton may be an integer with absolute magnitude greater than 1. However, a soliton with larger winding number would probably break up into solitons of winding number ± 1 to reduce the free energy, as we will discuss in Sect. 3.3.

2.2 Higher Dimensions

The concept of topological solitons can be generalized to higher dimensions. In 2D, a topological soliton for the xy model becomes a 1D linear structure, as illustrated in Fig. 2.3. This structure is normally called a wall, or a 2π-wall.

In Fig. 2.3a, the wall is straight. It separates two equivalent regions with $\hat{\boldsymbol{n}}(\boldsymbol{r})$ aligned along $\hat{\boldsymbol{x}}$. Between these uniform regions, across the soliton line, $\hat{\boldsymbol{n}}(\boldsymbol{r})$ rotates through an angle of 2π, and hence the winding number is 1. In Fig. 2.3b, the wall is curved. Again, it separates two equivalent regions with $\hat{\boldsymbol{n}}(\boldsymbol{r})$ along $\hat{\boldsymbol{x}}$, and the winding number is 1 as we cross the line in any direction.

A soliton line can terminate inside of a system. Its endpoint must be a topological defect, with topological charge equal to the winding number of the soliton line (with a positive or negative sign). As an example, Fig. 2.3c shows a soliton line

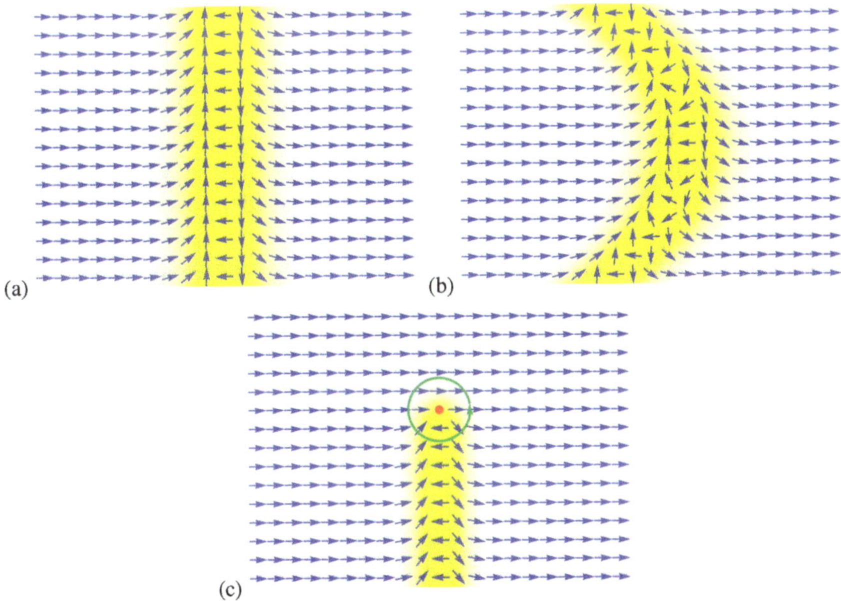

Fig. 2.3 Configurations of $\hat{\boldsymbol{n}}(\boldsymbol{r})$ with topological solitons in 2D. (**a**) Straight soliton line. (**b**) Curved soliton line. (**c**) Soliton line that terminates inside the system, with a defect of topological charge $+1$ at the endpoint

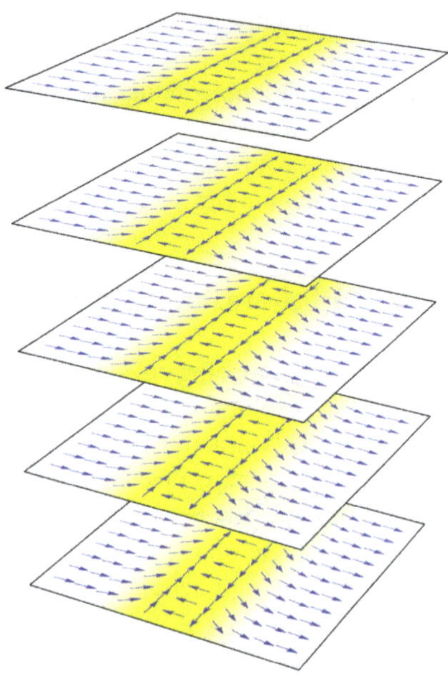

Fig. 2.4 Configuration of $\hat{\boldsymbol{n}}(\boldsymbol{r})$ with a topological soliton in 3D. In this example, the soliton is a straight, vertical wall

with winding number 1, which terminates with a topological defect of charge $+1$. (This defect looks somewhat different from the examples in Fig. 1.4a–d, because here the rotation of $\hat{\boldsymbol{n}}(\boldsymbol{r})$ is concentrated rather than spread out. Even so, we can tell that it has topological charge of $+1$ by integrating the increments of angle θ around the green loop that encloses the defect.)

In 3D, a topological soliton for the xy model becomes a 2D planar structure, as illustrated in Fig. 2.4. This structure is again called a wall, or a 2π-wall, and it separates two equivalent regions with $\hat{\boldsymbol{n}}(\boldsymbol{r})$ aligned along $\hat{\boldsymbol{x}}$. The wall may be either straight or curved. It may terminate inside of a system, and the edge will be a defect line.

In general, suppose the dimension of space is d_{space}, and the dimension of the measuring surface is d_{measure}, and we want to determine the dimension of solitons d_{soliton}. The soliton extends across the measuring surface, so it fills up all the dimensions of space except the dimensions of the measuring surface. Hence, the dimension of solitons becomes

$$d_{\text{soliton}} = d_{\text{space}} - d_{\text{measure}}. \tag{2.1}$$

This equation can be compared with Eq. (1.10) for topological defects.

2.3 Comparison of Defects and Solitons

We have seen that defects and solitons are both topological features of the xy model. They are both characterized by a winding number q, which indicates how $\hat{\boldsymbol{n}}(\boldsymbol{r})$ rotates as we move along a 1D measuring surface—the green path or loop in the figures. Hence, they are both described by the mathematical expression $\pi_1(S^1) = \mathbb{Z}$.

Based on these similarities, some researchers use the term "defects" for both defects and solitons. Readers should be prepared to see them both labeled as "defects" in the scientific literature. We believe that this terminology is misleading, and we do not recommend it. Topological defects and solitons are actually quite different from each other, and it is appropriate to identify them by two different names. Table 2.1 summarizes the main differences between them.

The last row of the table requires a special comment. A topological defect must have *long-range* effects on $\hat{\boldsymbol{n}}(\boldsymbol{r})$ in the surrounding system, because it can be detected by a large integration loop, far from the defect. As a consequence, if a system has many defects of the same sign, their effects add up to grow larger and larger. For example, Fig. 2.5a shows a system with four defects, each of topological charge $q = +1$. We can see that $\hat{\boldsymbol{n}}(\boldsymbol{r})$ becomes highly distorted away from the defects, with four full rotations as we go around the large green loop. By comparison, if half of the defects have the opposite topological charge $q = -1$, as in Fig. 2.5b, the long-range effects cancel and $\hat{\boldsymbol{n}}(\boldsymbol{r})$ becomes approximately uniform away from the defects. Hence, defects normally occur near other defects of the opposite charge, so that the total system has zero topological charge.

By contrast, a topological soliton has only *short-range* effects on $\hat{\boldsymbol{n}}(\boldsymbol{r})$, over a length scale comparable to the width ξ. Beyond that length scale, $\hat{\boldsymbol{n}}(\boldsymbol{r})$ returns to its background orientation (which is $\hat{\boldsymbol{x}}$ in the examples here). Hence, at longer length scales, it is impossible to detect the presence of the soliton. For that reason, if a system has many solitons with the same winding number, their effects do not add up. As an example, Fig. 2.5c shows a system with four solitons, each of winding number $q = +1$. Figure 2.5d shows a system with two solitons of $q = +1$ and two solitons of $q = -1$. In both cases, the solitons are almost independent of each

Table 2.1 Comparison of topological defects and solitons for the xy model

	Defects	Solitons
Singularity	*Singularity* where $\hat{\boldsymbol{n}}(\boldsymbol{r})$ is undefined	*No singularity*, $\hat{\boldsymbol{n}}(\boldsymbol{r})$ is well-defined everywhere
Measuring surface	Integrate along 1D loop *around* defect	Integrate along 1D path *through* soliton
Dimensionality	$d_{\text{defect}} = d_{\text{space}} - d_{\text{measure}} - 1$ Example: Point-like defect for 2D space, 1D loop	$d_{\text{soliton}} = d_{\text{space}} - d_{\text{measure}}$ Example: Point-like soliton for 1D space, 1D path
Long-range effects	Defect has *long-range* effects on $\hat{\boldsymbol{n}}(\boldsymbol{r})$	Soliton has *short-range* effects on $\hat{\boldsymbol{n}}(\boldsymbol{r})$

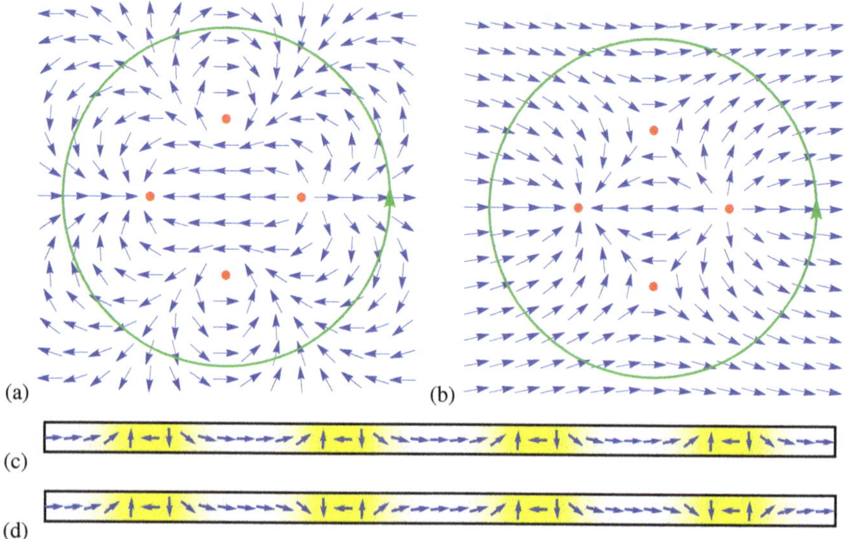

Fig. 2.5 (**a**) Four topological defects of $q = +1$, leading to large distortions in $\hat{\boldsymbol{n}}(\boldsymbol{r})$ at long range. (**b**) Two topological defects of $q = +1$, and two of $q = -1$, giving a uniform $\hat{\boldsymbol{n}}(\boldsymbol{r})$ at long range. (**c**) Four topological solitons of $q = +1$, with background orientation of $\hat{\boldsymbol{n}}(\boldsymbol{r})$ between the solitons. (**d**) Two topological solitons of $q = +1$, and two of $q = -1$, again with background orientation of $\hat{\boldsymbol{n}}(\boldsymbol{r})$ between the solitons

other, with the background orientation of $\hat{\boldsymbol{n}}(\boldsymbol{r})$ between them. Hence, a system does not need to make a total winding number of zero by including solitons of opposite winding number; the total may be positive or negative.

3 Free Energy

In the first two chapters, we discussed the *topological* properties of defects and solitons. For that discussion, we classified configurations of $\hat{\boldsymbol{n}}(\boldsymbol{r})$ based on the topological charge or winding number. Now, we need to consider the *physical* properties of these topological objects. In this chapter, we begin with the free energy.

Before going on, we should make a brief comment about terminology: In the literature on liquid crystals, it is conventional to use the term *free energy*. We have adopted that convention in this book, because it recognizes that the physical effects discussed here include both energetic and entropic contributions, and hence depend on temperature. In other subfields of physics, researchers may normally use the terms *energy* or *Hamiltonian* or *effective Hamiltonian*. For the purposes of this book, all of those terms are equivalent. We will make some further comments on energy and free energy in Sect. 4.2.2.

3.1 Free Energy of Defect

Let us consider a 2D system with a topological defect of charge q at the origin, as in Fig. 1.4. Everywhere around the defect, there is a well-defined orientation $\hat{\boldsymbol{n}}$, which depends on position $\boldsymbol{r}$. This spatial variation must cost free energy; a state with varying $\hat{\boldsymbol{n}}(\boldsymbol{r})$ has a higher free energy than a state with uniform $\hat{\boldsymbol{n}}$. How much higher? The simplest model just assumes that the free energy density is proportional to the square of the spatial gradient $\boldsymbol{\nabla}\hat{\boldsymbol{n}}$. Hence, the total free energy is the integral

$$\begin{aligned} F &= \frac{1}{2}K\int d\boldsymbol{r}|\boldsymbol{\nabla}\hat{\boldsymbol{n}}|^2 = \frac{1}{2}K\int d\boldsymbol{r}\left[(\boldsymbol{\nabla}\cdot\hat{\boldsymbol{n}})^2 + |\boldsymbol{\nabla}\times\hat{\boldsymbol{n}}|^2\right] \\ &= \frac{1}{2}K\int d\boldsymbol{r}(\partial_i n_j)(\partial_i n_j) = \frac{1}{2}K\int d\boldsymbol{r}|\boldsymbol{\nabla}\theta|^2, \end{aligned} \tag{3.1}$$

J. V. Selinger, *Introduction to Topological Defects and Solitons*, Lecture Notes in Physics 1032, https://doi.org/10.1007/978-3-031-70200-6_3

integrated over both dimensions of the system. The next-to-last expression uses tensor notation, which is implicitly summed over the indices i and j. The last expression uses the relationship $\hat{\boldsymbol{n}}(\boldsymbol{r}) = (\cos\theta(\boldsymbol{r}), \sin\theta(\boldsymbol{r}))$ between the unit vector $\hat{\boldsymbol{n}}(\boldsymbol{r})$ and the angle $\theta(\boldsymbol{r})$.

Equation (3.1) is called an elastic free energy, because it is analogous to the familiar potential energy $U = \frac{1}{2}kx^2$ for an elastic solid that follows Hooke's law. The coefficient K is called the elastic constant. In the liquid-crystal literature, the elastic free energy is more specifically called the Frank (or Oseen-Frank) free energy, with the approximation of a single Frank constant K.

The free energy of Eq. (3.1) is not the most general possible free energy. One generalization might be to include different coefficients for $(\nabla \cdot \hat{\boldsymbol{n}})^2$ and $|\nabla \times \hat{\boldsymbol{n}}|^2$, as we will discuss in Sect. 7.3. In the liquid-crystal literature, these terms are called splay and bend, respectively, for a 2D system.[1] However, for now, we will just use the simplest form of Eq. (3.1).

The question is now: Of all possible $\hat{\boldsymbol{n}}(\boldsymbol{r})$ configurations around the defect, what configuration gives the lowest free energy? This is a variational calculus problem.[2] Following the method of variational calculus, we calculate the functional derivative of the elastic free energy with respect to the angle $\theta(\boldsymbol{r})$, and set it equal to zero, which gives

$$\frac{\delta F}{\delta\theta(\boldsymbol{r})} = -K\nabla^2\theta = 0. \tag{3.2}$$

That equation is called the Euler-Lagrange equation for the problem. In this problem, it is Laplace's equation. It is most conveniently solved by converting from Cartesian coordinates (x, y) to polar coordinates (r, ϕ), with

$$x = r\cos\phi, \quad y = r\sin\phi, \quad r = \sqrt{x^2 + y^2}, \quad \phi = \tan^{-1}\left(\frac{y}{x}\right). \tag{3.3}$$

In polar coordinates, the Euler-Lagrange equation becomes

$$-K\left[\frac{\partial^2\theta}{\partial r^2} + \frac{1}{r}\frac{\partial\theta}{\partial r} + \frac{1}{r^2}\frac{\partial^2\theta}{\partial\phi^2}\right] = 0. \tag{3.4}$$

[1] For our discussion of the general Oseen-Frank free energy, see J. V. Selinger, "Interpretation of Saddle-Splay and the Oseen-Frank Free Energy in Liquid Crystals," *Liq. Cryst. Rev.* **6**, 129 (2018).

[2] For a quick physical introduction to variational calculus, see our previous book J. V. Selinger, *Introduction to the Theory of Soft Matter: From Ideal Gases to Liquid Crystals* (Springer, 2016), Chapter 5. For a more mathematical introduction, see M. Stone and P. Goldbart, *Mathematics for Physics: A Guided Tour for Graduate Students* (Cambridge, 2009); G. B. Arfken, H. J. Weber, and F. E. Harris, *Mathematical Methods for Physicists: A Comprehensive Guide*, 7th edition (Elsevier, 2013).

A solution of this partial differential equation, subject to the boundary condition associated with the defect, is

$$\theta(r, \phi) = q\phi + \theta_0, \tag{3.5}$$

where q is the topological charge of the defect and θ_0 is a constant angle. The physical significance of θ_0 will be discussed in Sect. 6.1. If we move around any loop about the defect, the coordinate ϕ increases by 2π, so the angle θ increases by $2\pi q$. Hence, $\hat{\boldsymbol{n}}$ rotates q times through a full circle in the counter-clockwise direction, which is the definition of the topological charge.

In Cartesian coordinates, the solution for θ becomes

$$\theta(x, y) = q \tan^{-1}\left(\frac{y}{x}\right) + \theta_0. \tag{3.6}$$

A plot of this solution for $\theta(x, y)$ is shown in Fig. 3.1. The corresponding unit vector has the Cartesian components $\hat{\boldsymbol{n}}(x, y) = (\cos\theta(x, y), \sin\theta(x, y))$. Some examples of this solution, with different values of q and θ_0, are illustrated in Fig. 1.4. If the defect is located away from the origin, at a point (x_d, y_d), then we can easily shift the solution to

$$\theta(x, y) = q \tan^{-1}\left(\frac{y - y_d}{x - x_d}\right) + \theta_0, \tag{3.7}$$

again with $\hat{\boldsymbol{n}}(x, y) = (\cos\theta(x, y), \sin\theta(x, y))$.

The next question is: How much free energy is associated with the $\hat{\boldsymbol{n}}(\boldsymbol{r})$ configuration that we have calculated? To answer that question, we take the gradient of our solution,

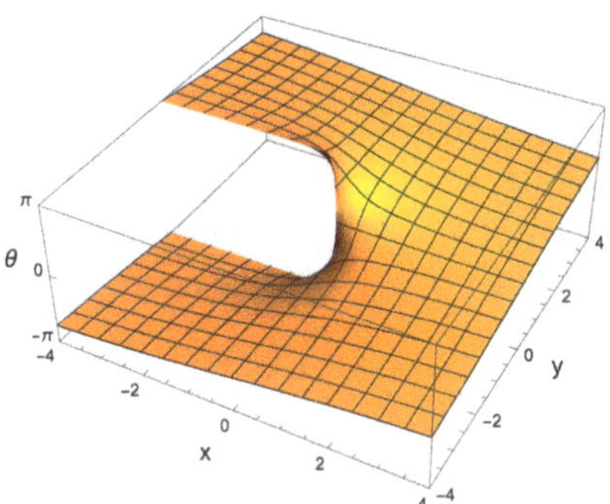

Fig. 3.1 Plot of the solution $\theta(x, y)$ for a single defect at the origin, from Eq. (3.6), with $q = 1$ and $\theta_0 = 0$. Along the negative x-axis, θ is discontinuous but it jumps by 2π, which has no effect on the vector $\hat{\boldsymbol{n}} = (\cos\theta, \sin\theta)$. Hence, this solution is really only singular at the origin (The mathematical term for the discontinuity with no physical consequence is a *branch cut*.)

$$\nabla\theta = \frac{\partial\theta}{\partial r}\hat{\boldsymbol{r}} + \frac{1}{r}\frac{\partial\theta}{\partial\phi}\hat{\boldsymbol{\phi}} = \frac{q}{r}\hat{\boldsymbol{\phi}}, \quad |\nabla\theta|^2 = \frac{q^2}{r^2}, \tag{3.8}$$

then substitute it back into the free energy of Eq. (3.1), and integrate over the entire system in polar coordinates. For an infinite system, this integration gives

$$F = \frac{1}{2}K\int_0^\infty r dr\int_0^{2\pi} d\phi\left(\frac{q^2}{r^2}\right) = \pi K q^2\int_0^\infty \frac{dr}{r} = \pi K q^2 \log r\Big|_0^\infty, \tag{3.9}$$

where log means natural logarithm, base e. Unfortunately, this integral does not converge at either the lower limit $r \to 0$ or the upper limit $r \to \infty$. Let us deal with each of these divergences separately.

The divergence at $r \to 0$ is called an *ultraviolet* divergence, because it occurs on short length scales, just as ultraviolet light has a shorter wavelength than visible light. It occurs because we are applying the free energy of Eq. (3.1) down to microscopic length scales around the singularity. Actually, this free energy cannot apply there, for several reasons. First, because $\hat{\boldsymbol{n}}(\boldsymbol{r})$ varies so rapidly as a function of position around the singularity, the free energy should not neglect higher-order terms. Rather, it should include higher derivatives of $\hat{\boldsymbol{n}}(\boldsymbol{r})$, as well as higher powers of first derivatives. Even more importantly, the magnitude of the xy order parameter may change near the singularity, so that the system can no longer be described by a unit vector $\hat{\boldsymbol{n}}(\boldsymbol{r})$. We will discuss that issue in Chap. 5. For now, we can just say that the free energy applies down to a minimum length scale $r_{\mathrm{core}} > 0$, which can be regarded as a defect core radius. In the integral of Eq. (3.9), the lower limit should be r_{core} rather than 0.

The divergence at $r \to \infty$ is called an *infrared* divergence, because it occurs on long length scales, just as infrared light has a longer wavelength than visible light. Unlike the ultraviolet divergence, the infrared divergence is a real physical effect, because the free energy of Eq. (3.1) really applies at long length scales. It occurs because the defect creates long-range distortions of $\hat{\boldsymbol{n}}(\boldsymbol{r})$, which decay very slowly with distance, with $|\nabla\theta| \propto 1/r$. This decay is so slow that the free energy associated with these distortions does not converge. If we had a single defect in an infinite system, the free energy would diverge. To address this issue, we note that the system is not really infinite. Instead, suppose that the sample is a disk with radius R. In that case, in the integral of Eq. (3.9), the upper limit should be R rather than ∞.

With these two changes, the integrated free energy becomes

$$F = \pi K q^2 \log r\Big|_{r_{\mathrm{core}}}^{R} = \pi K q^2 \log\left(\frac{R}{r_{\mathrm{core}}}\right). \tag{3.10}$$

We should make several remarks about this result:

First, we consider this result as the free energy of the defect, because it shows the extra free energy of a system with a defect, compared with a system with no defect. However, the free energy density is not concentrated inside the singularity. Rather, it is spread out over the entire system. It is highest near the singularity, but it decays very slowly as a function of distance away from the singularity. The result shows us that the singularity is accompanied by a very long-range distortion of $\hat{\boldsymbol{n}}(\boldsymbol{r})$, and the free energy is associated with the entire distortion.

Second, the free energy is proportional to the topological charge q squared. For that reason, a defect of charge 2 has a higher free energy than two defects of charge 1. Any defect with a high charge will tend to break up into many defects with the minimum possible charge, which is ± 1, in order to reduce the free energy. Hence, we should expect to observe only defects of the minimum charge; higher-charge defects will be very rare.

Third, the free energy depends on the system size R and the core radius r_{core}, but the dependence is only logarithmic, which is a very weak dependence. For example, if the system size is $R = 1$ cm, and the core radius is $r_{\text{core}} = 1$ nm, then $\log(R/r_{\text{core}}) = \log(10^7) \approx 16$, which is only a modest factor. If the system size is increased greatly to $R = 1$ m, then $\log(R/r_{\text{core}}) = \log(10^9) \approx 21$, which is not much larger. Hence, the result is not very sensitive to R and r_{core}.

Fourth, the free energy is also not very sensitive to the shape of the system. For example, suppose we have a defect in the middle of a rectangular sample of size $L_x \times L_y$, in the limit of $L_y \to \infty$. In this case, the free energy is most conveniently calculated in Cartesian coordinates as

$$F = \frac{1}{2} K(4) \int_{r_{\text{core}}}^{L_x/2} dx \int_0^{\infty} dy \frac{1}{x^2 + y^2} = \pi K q^2 \log\left(\frac{L_x}{r_{\text{core}}}\right). \tag{3.11}$$

This result is very similar to the free energy of Eq. (3.10), with R replaced by L_x, the shortest distance from the defect to the boundary. We have treated the lower cutoff r_{core} differently in Eqs. (3.10) and (3.11), but it does not really matter because the dependence on r_{core} is only logarithmic.

Fifth, some authors write the free energy as $F = \pi K q^2 \log(R/r_{\text{core}}) + F_{\text{core}}$, where F_{core} is the free energy of the defect core. This parameter describes the unknown free energy over the range of length scales $0 < r < r_{\text{core}}$. Here, we omit that parameter, and we just absorb the core free energy into the lower cutoff r_{core}. We will return to this topic with a more detailed theory in Chap. 5.

As a brief aside, we can see how the same calculation can be adapted for a straight defect line in a 3D system, as in Fig. 1.6a. In this case, we would modify the free energy of Eq. (3.1) to

$$F = \frac{1}{2} K \int d\boldsymbol{r} |\nabla \hat{\boldsymbol{n}}|^2 = \frac{1}{2} K \int d\boldsymbol{r} |\nabla \theta|^2, \tag{3.12}$$

integrated over all three dimensions. Careful readers may notice that the elastic constant has different units in Eqs. (3.1) and (3.12), because of the different dimensions of the integral. In this book, we implicitly assume that the elastic constant has units corresponding to the dimensionality of our system: $[K] =$ (energy)(length) in 1D, energy in 2D, energy/length in 3D. For a straight defect line in the z-direction, $\theta(\boldsymbol{r})$ depends only on x and y, not on z. Hence, the 2D calculation for $\theta(\boldsymbol{r})$ can be repeated without change. When we put that solution back into the free energy of Eq. (3.12), we find that $F = \pi K q^2 \log(R/r_{\text{core}})$ is the free energy *per length in the z-direction.* In other words, it is the line tension.

3.2 Interaction Between Defects

Let us now generalize the calculation of the previous section to a system with *two* defects.[3] In particular, suppose there is one defect of topological charge q_1 at position $\boldsymbol{r}_1 = (x_1, y_1)$, and another defect of topological charge q_2 at position $\boldsymbol{r}_2 = (x_2, y_2)$. As a first step, we must ask: Of all possible $\hat{\boldsymbol{n}}(\boldsymbol{r})$ configurations around the two defects, what configuration gives the lowest free energy? In other words, we must solve the Euler-Lagrange equation (3.2) for two defects. This calculation is straightforward, because the Euler-Lagrange equation is linear in $\theta(\boldsymbol{r})$. Linear differential equations satisfy the principle of *superposition*, meaning that any sum of solutions is also a solution. Hence, we just add together solutions of the form (3.7) corresponding to each defect, in Cartesian coordinates, to obtain

$$\begin{aligned} \theta(x, y) &= \left[q_1 \tan^{-1}\left(\frac{y - y_1}{x - x_1}\right) + \theta_1\right] + \left[q_2 \tan^{-1}\left(\frac{y - y_2}{x - x_2}\right) + \theta_2\right] \\ &= q_1 \tan^{-1}\left(\frac{y - y_1}{x - x_1}\right) + q_2 \tan^{-1}\left(\frac{y - y_2}{x - x_2}\right) + \theta_0, \end{aligned} \tag{3.13}$$

still with $\hat{\boldsymbol{n}}(x, y) = (\cos\theta(x, y), \sin\theta(x, y))$. Examples of this combined solution for $\theta(x, y)$ are plotted in Fig. 3.2, and the corresponding results for $\hat{\boldsymbol{n}}(x, y)$ are illustrated in Fig. 1.5. Note that we have combined the constant terms $\theta_1 + \theta_2$ into a single constant θ_0; the physical significance of this combination will be discussed in Sect. 6.1.

Next, we must calculate the total free energy associated with this solution for $\theta(\boldsymbol{r})$. To simplify the calculation, suppose the defect positions are $\boldsymbol{r}_1 = (x_d, 0)$ and $\boldsymbol{r}_2 = (-x_d, 0)$, so that the distance between them is $r_{12} = 2x_d$. The gradient of our solution becomes

$$\nabla\theta = \frac{q_1[-y\hat{\boldsymbol{x}} + (x - x_d)\hat{\boldsymbol{y}}]}{(x - x_d)^2 + y^2} + \frac{q_2[-y\hat{\boldsymbol{x}} + (x + x_d)\hat{\boldsymbol{y}}]}{(x + x_d)^2 + y^2}, \tag{3.14}$$

[3] This calculation, like all the calculations in this chapter, is considered classical work in the theory of topological defects and solitons. We do not know the history well enough to identify all of the original references. As far as we know, this calculation was first presented by C. M. Dafermos, "Disinclinations in Liquid Crystals," *Q. J. Mech. Appl. Math.* **23**, 49 (1970). (This type of defect was previously spelled as *disinclination*, and the spelling has since been shortened to *disclination*.)

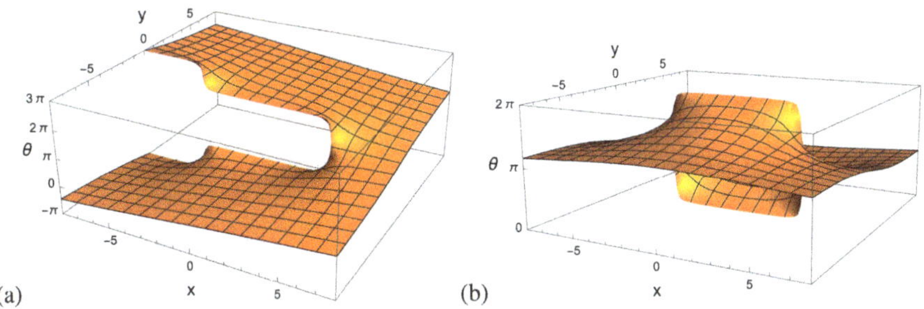

Fig. 3.2 Plots of the solution $\theta(x, y)$ for two defects, from Eq. (3.13). (**a**) Example with $q_1 = q_2 = 1$ and $\theta_0 = \pi$. (**b**) Example with $q_1 = 1$, $q_2 = -1$, and $\theta_0 = \pi$. These examples correspond to the $\hat{\boldsymbol{n}}(x, y)$ plots in Fig. 1.5a,b, respectively

$$|\nabla\theta|^2 = \frac{q_1^2}{(x-x_d)^2+y^2} + \frac{q_2^2}{(x+x_d)^2+y^2} + \frac{2q_1q_2(x^2-x_d^2+y^2)}{[(x-x_d)^2+y^2][(x+x_d)^2+y^2]}.$$

Hence, the free energy integral has terms proportional to q_1^2, q_2^2, and q_1q_2. The q_1^2 term is just the free energy of a single defect of charge q_1 at position $\boldsymbol{r}_1$. If we calculate this integral using a lower cutoff r_{core} and an upper cutoff L_x, as in Eq. (3.11), and assume $L_x \gg x_d \gg r_{\text{core}}$, it becomes $F_1 = \pi K q_1^2 \log(L_x/r_{\text{core}})$. Likewise, the q_2^2 term is the free energy of a single defect of charge q_2 at position $\boldsymbol{r}_2$, and it becomes $F_2 = \pi K q_2^2 \log(L_x/r_{\text{core}})$. The cross term can be calculated as

$$\begin{aligned} F_{\text{int}} &= \frac{1}{2}K(4)\int_0^{L_x/2} dx \int_0^{\infty} dy \frac{2q_1q_2(x^2-x_d^2+y^2)}{[(x-x_d)^2+y^2][(x+x_d)^2+y^2]} \\ &= 2\pi K q_1 q_2 \log\left(\frac{L_x}{2x_d}\right) = 2\pi K q_1 q_2 \log\left(\frac{L_x}{r_{12}}\right). \end{aligned} \tag{3.15}$$

The result in Eq. (3.15) shows the extra free energy associated with two defects separated by a distance r_{12}, compared with two isolated defects. It can be interpreted as an *interaction* between the two defects. The interaction free energy is not concentrated inside the defects themselves; rather, it is spread out over the entire system. We might say that the interaction is mediated by the $\hat{\boldsymbol{n}}(\boldsymbol{r})$ distortion around the defects.

Because the interaction free energy depends on the distance r_{12}, it generates a force

$$f = -\frac{\partial F}{\partial r_{12}} = \frac{2\pi K q_1 q_2}{r_{12}}. \tag{3.16}$$

This force is quite analogous to the force between electric charges in Coulomb's law. It is repulsive if q_1 and q_2 have the same sign, and attractive if they have opposite signs. Even the distance dependence is the same as Coulomb's law in a

2D universe.[4] Note that the force scales as $1/r_{12}$, and the free energy as $\log(r_{12})$, and these functions both depend very weakly on distance. Hence, defects experience a *long-range* interaction, which tends to pull opposite-sign defects together so that they annihilate each other, and tends to push like-sign defects apart.

This calculation can easily be adapted for the interaction of two straight defect lines, both running in the z-direction, in a 3D system. In that case, Eq. (3.15) is still valid, but F_{int} must be interpreted as the interaction free energy per length in the z-direction. Likewise, Eq. (3.16) is still valid, but f is the force per length in the z-direction. The units of energy per length and force per length are correct, considering that the units of K have changed, as discussed at the end of Sect. 3.1.

3.3 Free Energy of Soliton

We now consider the free energy associated with a topological soliton. We will see that it is quite different from the free energy of a topological defect.

In all of the examples of topological solitons in Chap. 2, $\hat{\boldsymbol{n}}(\boldsymbol{r})$ has a characteristic background orientation $\hat{\boldsymbol{x}}$. It only deviates from that background orientation in a small region. When we model the free energy, we must include some physical mechanism to align $\hat{\boldsymbol{n}}(\boldsymbol{r})$ in the background orientation. If there were no such mechanism, then the variations of $\hat{\boldsymbol{n}}(\boldsymbol{r})$ would be spread out rather than concentrated.

The simplest mechanism to align $\hat{\boldsymbol{n}}(\boldsymbol{r})$ in the background orientation is a magnetic field. Suppose the system has a uniform magnetic field $\boldsymbol{B} = B\hat{\boldsymbol{x}}$. It couples linearly to the magnetic moment density $\boldsymbol{\mu} = \mu\hat{\boldsymbol{n}}(\boldsymbol{r})$ through a term of $-\boldsymbol{\mu}\cdot\boldsymbol{B} = -\mu\hat{\boldsymbol{n}}\cdot\boldsymbol{B}$, and hence favors the alignment $\hat{\boldsymbol{n}}(\boldsymbol{r}) = \hat{\boldsymbol{x}}$. The coefficient μ is a positive constant, with units of $[\mu B] = \text{energy}/\text{length}$ in 1D, energy/length^2 in 2D, and energy/length^3 in 3D. Combining the elastic and magnetic contributions, the free energy is

$$
\begin{aligned}
F &= \int_{-\infty}^{\infty} dx \left[\frac{1}{2}K|\nabla\hat{\boldsymbol{n}}|^2 - \mu\hat{\boldsymbol{n}}\cdot\boldsymbol{B} + \mu B\right] \\
&= \int_{-\infty}^{\infty} dx \left[\frac{1}{2}K\left(\frac{d\theta}{dx}\right)^2 + \mu B(1-\cos\theta)\right].
\end{aligned}
\tag{3.17}
$$

Here we have added an unimportant constant μB, so that $F = 0$ for a uniform system with $\hat{\boldsymbol{n}} = \hat{\boldsymbol{x}}$.

Let us make a model for a single soliton in a 1D system, as in Fig. 2.1. For a soliton of winding number $q = +1$, the orientation begins at $\hat{\boldsymbol{n}} = \hat{\boldsymbol{x}}$ for large negative x, then it rotates through a full circle in the counter-clockwise direction

[4] In a 2D universe (unlike the actual 3D universe), Gauss's law would give an electric field proportional to $1/r$. Hence, the electrostatic force in Coulomb's law would depend on distance as $1/r$.

as x increases, and then it returns to $\hat{\boldsymbol{n}} = \hat{\boldsymbol{x}}$ for large positive x. Equivalently, the angle begins at $\theta = 0$ for large negative x, then θ increases as x increases, then $\theta = 2\pi$ for large positive x. Hence, we want to minimize the free energy subject to the boundary conditions that $\theta(-\infty) = 0$ and $\theta(\infty) = 2\pi$. Following the methods of variational calculus, we derive the Euler-Lagrange equation

$$-K\frac{d^2\theta}{dx^2} + \mu B \sin\theta = 0, \tag{3.18}$$

or equivalently

$$\frac{d^2\theta}{dx^2} = \frac{1}{\xi^2}\sin\theta, \text{ with } \xi = \sqrt{\frac{K}{\mu B}}. \tag{3.19}$$

This is a standard equation of mathematical physics, known as the *sine-Gordon equation*.[5] It is nonlinear in the angle $\theta(x)$, and hence it can be difficult to solve. With our current boundary conditions, it has the exact solution

$$\theta(x) = \pi + 4\tan^{-1}\left[\tanh\left(\frac{x - x_0}{2\xi}\right)\right]. \tag{3.20}$$

This solution for $\theta(x)$ is plotted in Fig. 3.3, and the corresponding results for $\hat{\boldsymbol{n}}(x)$ are illustrated in Fig. 2.1. Note that the soliton has the form of a kink, which separates two regions of the favorable orientation $\theta = 0$ and 2π, where $\hat{\boldsymbol{n}}(x)$ is aligned with the magnetic field. The parameter x_0 represents the location of the center of the kink, which may be anywhere in the infinite system. The parameter ξ shows the width of the kink. From Eq. (3.19), we can see that the kink is narrow when the elastic constant K is small and the magnetic field B is large. Conversely, the kink is spread-out when K is large and B is small.

We next ask: How much free energy is associated with the soliton? We put our solution for $\theta(x)$ back into the free energy of Eq. (3.17), and integrate over x. The integral converges because the $\hat{\boldsymbol{n}}(x)$ distortions associated with the soliton are short-ranged; the elastic and magnetic terms both decay exponentially as $x \to \pm\infty$. The result is

$$F = 8\sqrt{K\mu B}. \tag{3.21}$$

[5] The name *sine-Gordon equation* was originally a pun, because the equation looks like the *Klein-Gordon equation* $d^2\theta/dx^2 = (1/\xi^2)\theta$, but with $\sin\theta$ in place of θ. Now it is a well-accepted name, not a joke. Here, we have the time-independent sine-Gordon equation, not the full time-dependent equation.

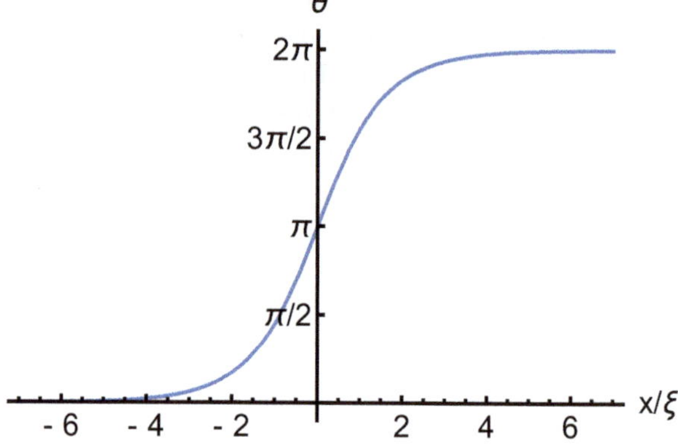

Fig. 3.3 Plot of the solution $\theta(x)$ for a single soliton centered at $x_0 = 0$, from Eq. (3.20). This example corresponds to the $\hat{\boldsymbol{n}}(x)$ plots in Fig. 2.1

Thus, we see that the elastic constant K and the magnetic field alignment μB are the two essential pieces needed to make a soliton. The ratio of these parameters gives the soliton width ξ, and their geometric mean gives the soliton free energy F.

What if we want a soliton of winding number $q = -1$? The solution for $\theta(x)$ is just the negative of Eq. (3.20), modulo 2π. Hence, we can write $\theta(x) = \pi - 4\tan^{-1}[\tanh(x - x_0)/(2\xi)]$. The width ξ and free energy F are exactly the same as calculated above.

What if we want a soliton of winding number $q = +2$ (or any other integer greater than $+1$ or less than -1)? That is a more difficult problem. We *cannot* just multiply our previous solution by 2, because the sine-Gordon equation is nonlinear; it does not satisfy the superposition principle. If we try to minimize the free energy with boundary conditions that $\theta(-\infty) = 0$ and $\theta(\infty) = 2\pi$, then the system will make two solitons, each with winding number $q = +1$, and these two solitons will move apart. Hence, we must say that the solitons of higher winding number are permitted by topology, but they are not favored by free energy.

The 1D calculation for winding number $q = \pm 1$ can easily be adapted for a soliton in a higher-dimensional system, such as the soliton line in Fig. 2.3a or the soliton plane in Fig. 2.4. In those cases, $\theta(\boldsymbol{r})$ depends on only one coordinate x, and is independent of y and z. The same solution for $\theta(x)$ still applies, along with the same results for ξ and F. Now, ξ must be interpreted as the thickness of the soliton line or plane. Likewise, F must be interpreted as the free energy *per length* of a soliton line, or the free energy *per area* of a soliton plane. The units are correct, considering that the units of K and μB depend on the dimensionality of the system, as discussed above.

3.4 Interaction Between Solitons

Now that we have the free energy of a single soliton, let us determine the interaction between two solitons.

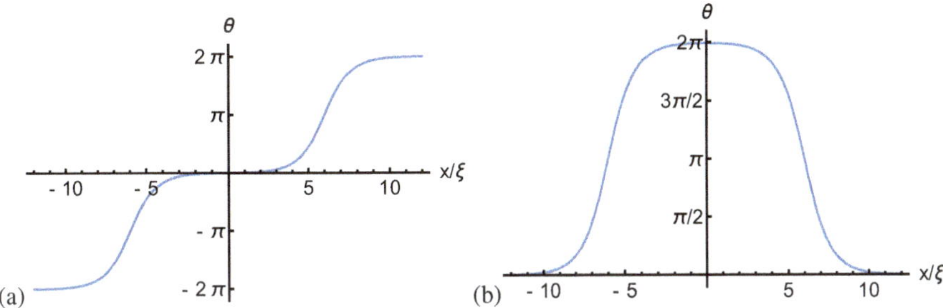

Fig. 3.4 Plot of the approximate solution $\theta(x)$ for a two solitons, from Eq. (3.22). (**a**) Winding numbers $q_1 = q_2 = +1$. (**b**) Winding numbers $q_1 = +1$, $q_2 = -1$. These examples corresponds to the $\hat{\boldsymbol{n}}(x)$ plots in Fig. 2.2

Suppose we have a 1D system with a soliton of winding number $q_1 = \pm 1$ at position x_1, and another soliton of winding number $q_2 = \pm 1$ at position x_2. Ideally, we would like to find the function $\theta(x)$ that minimizes the free energy with those two solitons, and then calculate the free energy with that $\theta(x)$. Unfortunately, this procedure cannot be done exactly, because the sine-Gordon equation is nonlinear.

Instead, we must use an approximation. Consider the sum of solutions for single solitons

$$\theta(x) = 4q_1 \tan^{-1}\left[\tanh\left(\frac{x - x_1}{2\xi}\right)\right] + 4q_2 \tan^{-1}\left[\tanh\left(\frac{x - x_2}{2\xi}\right)\right], \tag{3.22}$$

modulo 2π. It is approximately a solution of the sine-Gordon equation in the limit of $x_{12} = |x_1 - x_2| \gg \xi$. The error associated with this solution decreases as $e^{-x_{12}/\xi}$. Hence, we can use this approximate solution as long as the solitons are well-separated, recognizing that it breaks down when the solitons get close together. It is plotted in Fig. 3.4, and the corresponding results for $\hat{\boldsymbol{n}}(x)$ are shown in Fig. 2.2.

We now put this approximate solution into the free energy of Eq. (3.17), using the constraints $q_1 = \pm 1$ and $q_2 = \pm 1$, and integrate over x. The result is

$$F = 16\sqrt{K\mu B}\left[1 + 2q_1 q_2 e^{-x_{12}/\xi} + \ldots\right], \tag{3.23}$$

neglecting higher-order terms in the small parameter $e^{-x_{12}/\xi}$. In this result, the first term is the free energy of the two solitons, as calculated previously, and the second term is the interaction between the solitons. It generates a force

$$f = -\frac{\partial F}{\partial x_{12}} = 32\mu B q_1 q_2 e^{-x_{12}/\xi}. \tag{3.24}$$

Like the force between two defects, this force between two solitons is repulsive if the solitons have the same winding number, and attractive if they have opposite winding number. Unlike the force between two defects, the force between two solitons has only a short range. It decays exponentially as a function of separation, with a characteristic decay length of ξ, which is the width of the solitons. Hence, this calculation is another way of seeing the short-range properties of solitons.

The same calculation also works for a soliton in a higher-dimensional system. It gives the force per length acting on a soliton line, or the force per area acting on a soliton plane.

3.5 Possibility of Stable Defects or Solitons

So far, in all of the calculations in this chapter, we have considered situations in which defects or solitons are *metastable*. In other words, the defects or solitons are local minima of the free energy, satisfying the Euler-Lagrange equation, but they are not the global minima. The *stable states*, i.e. global minima of the free energy, have uniform $\hat{\boldsymbol{n}}(\boldsymbol{r})$ (aligned with the magnetic field $\boldsymbol{B}$, if it is present).

One might ask: Is there any situation in which defects or solitons can be stable states, lower in free energy than a uniform state? That is possible if the free energy includes some mechanism that actually favors gradients of $\hat{\boldsymbol{n}}(\boldsymbol{r})$. In this section, we will discuss a simple example that has stable solitons.[6] We will defer more complex examples to Chap. 12.

For this simple example of stable solitons, we consider a 1D system extended in the z-direction from $-\infty$ to $+\infty$, as shown in Fig. 3.5. The unit vector $\hat{\boldsymbol{n}}(\boldsymbol{r})$ is still in the (x, y) plane, and it depends on the coordinate z, so it can be written as

$$\hat{\boldsymbol{n}}(z) = (\cos\theta(z), \sin\theta(z), 0). \tag{3.25}$$

The free energy still has the elastic and magnetic terms discussed in the previous sections. In addition, suppose that the material is *chiral*; i.e. it does not have any reflection or inversion symmetry. In that case, the free energy can have an additional term of the form $D\hat{\boldsymbol{n}} \cdot \nabla \times \hat{\boldsymbol{n}}$. In the magnetic literature, this term is called a Dzyaloshinskii-Moriya interaction, which is a type of spin-orbit coupling that occurs in noncentrosymmetric materials. In the liquid-crystal literature, it is called a chiral term, which favors twist in cholesteric liquid crystals. The coefficient D indicates the strength of the asymmetry.

Combining all three terms, the total free energy becomes

$$F = \int_{-\infty}^{+\infty} dz \left[\frac{1}{2} K |\nabla \hat{\boldsymbol{n}}|^2 + \mu B - \mu \boldsymbol{B} \cdot \hat{\boldsymbol{n}} + D\hat{\boldsymbol{n}} \cdot \nabla \times \hat{\boldsymbol{n}} \right]$$

[6] This simple example is taken from the review article R. D. Kamien and J. V. Selinger, "Order and Frustration in Chiral Liquid Crystals," *J. Phys. Condens. Matter* **13**, R1 (2001), which also discusses more complex examples.

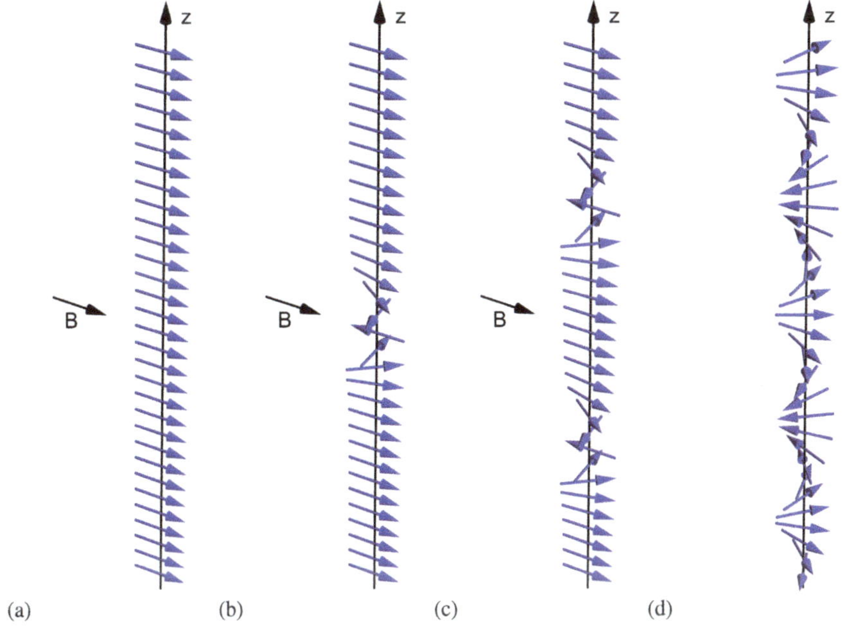

Fig. 3.5 Example of a system with stable solitons. (**a**) Uniform state with no solitons. (**b**) One soliton. (**c**) Two solitons. (**d**) Many solitons, leading to a helical modulation of $\hat{\boldsymbol{n}}(z)$

$$= \int_{-\infty}^{\infty} dz \left[\frac{1}{2} K \left(\frac{d\theta}{dz} \right)^2 + \mu B (1 - \cos\theta) - D \frac{d\theta}{dz} \right]. \tag{3.26}$$

If the magnetic field were $B = 0$, then the minimum could be easily calculated as $d\theta/dz = D/K$, so $\theta(z) = Dz/K +$ constant. This state has a helical modulation of $\hat{\boldsymbol{n}}(z)$, illustrated in Fig. 3.5d. It is the familiar structure of a helimagnet or cholesteric liquid crystal.

Here, we want to consider the different case of large B, so that $\hat{\boldsymbol{n}}(z)$ forms solitons. To solve that problem, we notice that the chiral term is a total derivative, which can be integrated exactly to give

$$F = \int_{-\infty}^{+\infty} dz \left[\frac{1}{2} K \left(\frac{d\theta}{dz} \right)^2 + \mu B (1 - \cos\theta) \right] - D \left[\theta(+\infty) - \theta(-\infty) \right]. \tag{3.27}$$

We can see that the chiral term depends only on the limits at $\pm\infty$, and it favors a difference between $\theta(+\infty)$ and $\theta(-\infty)$. The chiral term does not enter into the Euler-Lagrange equation, which is still the sine-Gordon equation (3.19) but with x replaced by z. Hence, solitons still have the same form $\theta(z)$ that was calculated in the previous sections.

Now we can compare the free energy of the uniform state (Fig. 3.5a) with the free energy of a single soliton with $q = +1$ (Fig. 3.5b). The uniform state has $\theta(z) = 0$ everywhere, and hence $F_{\text{uniform}} = 0$. For the single soliton, the K and B terms in the free energy give $8\sqrt{K\mu B}$, as calculated in Eq. (3.21). In addition, the D term gives an extra contribution of $-2\pi D$, because $\theta(z)$ increases by 2π from $\theta(-\infty)$ to $\theta(+\infty)$. Hence, the total free energy of a single soliton with $q = +1$ is

$$F_{\text{soliton}} = 8\sqrt{K\mu B} - 2\pi D. \tag{3.28}$$

The free energies are equal, $F_{\text{uniform}} = F_{\text{soliton}}$, at the threshold $8\sqrt{K\mu B} = 2\pi D$. If the chiral coefficient D is higher, or the magnetic field B is lower, then the single soliton is actually stable; it has a lower free energy than the uniform state.

What about a single soliton with $q = -1$? In that case, $\theta(z)$ decreases by 2π from $\theta(-\infty)$ to $\theta(+\infty)$, so the free energy is $F_{\text{soliton}} = 8\sqrt{K\mu B}+2\pi D$. Hence, the chiral coefficient D breaks the symmetry between positive and negative q. Winding number $q = +1$ is favored by $D > 0$, and winding number $q = -1$ is favored by $D < 0$.

Now we might ask: If the system can reduce its free energy by adding one soliton of $q = +1$, can it reduce its free energy further by adding two solitons of $q = +1$ (Fig. 3.5c)? Or even more solitons? For any system with more than one soliton, we must consider the exponential interaction between solitons. This unfavorable interaction energy competes with the favorable energy of a single soliton. The competition leads to an optimal spacing between solitons Δz, and hence an optimal density of solitons $1/\Delta z$. Right at the threshold $8\sqrt{K\mu B} = 2\pi D$, the optimal spacing is infinite, so that the lowest-free-energy state has just one soliton in the entire system. As we go beyond that threshold (increasing D or reducing B), the optimal spacing decreases, the optimal density increases, and the lowest-free-energy state has more and more solitons. Eventually, we reach a state where the spacing Δz is comparable to the width ξ, which is equivalent to the standard helimagnetic or cholesteric state (Fig. 3.5d).

This example demonstrates that solitons can be stable, with a free energy lower than the uniform state. The stability of solitons requires a combination of three physical effects:

1. A free energy benefit for certain spatial variations of the order parameter, which are allowed by the presence of solitons. In this example, it is provided by the chiral coupling D.
2. A free energy penalty for each soliton. In this example, it is provided by the soliton free energy $8\sqrt{K\mu B}$, arising from the elastic constant K and magnetic field B.
3. A repulsive interaction between solitons. In this example, it is provided by the exponential interaction calculated in Sect. 3.4, which also arises from K and B.

If the free energy benefit #1 exceeds the free energy penalty #2, then a single soliton has a lower free energy than the uniform state. In that case, the favored spacing between solitons is determined by the balance between the net benefit of a soliton and the interaction between solitons #3.

This combination of effects occurs in many other physical contexts, leading to stable solitons or even to stable topological defects. We will discuss some of these more complex cases in Chap. 12.

Dynamics and Statistical Mechanics 4

In the previous chapter, we discussed the free energy of topological defects and solitons. Here, we consider other physical properties: the dynamics of defects and solitons, and the role of defects in statistical mechanics.

4.1 Dynamics of Defects and Solitons

In this section, we want to begin with a model for the dynamics of the order parameter, and use this model to predict the dynamics of topological defects and solitons. As a first step, we must choose a model for dynamics of the order parameter. For simplicity, we choose overdamped dynamics, which relaxes toward a minimum of the free energy. Other types of dynamics are also possible, but they are beyond the scope of this book.[1]

We can model overdamped dynamics by analogy with the motion of a particle in a viscous fluid. Suppose we have a particle of mass m, with position $x(t)$, velocity $\dot{x}(t)$, and acceleration $\ddot{x}(t)$, moving in a potential $U(x)$. It experiences two forces: a force from the potential $f_{\text{potential}} = -\partial U/\partial x$ and a force from drag. The force from drag can be determined using the concept of a Rayleigh dissipation function D, which represents the rate of energy dissipation. For motion in a viscous fluid, the dissipation function is $D = \frac{1}{2}b\dot{x}^2$, where b is the drag coefficient, so the drag force is $f_{\text{drag}} = -\partial D/\partial \dot{x} = -b\dot{x}$. Newton's second law then gives $m\ddot{x} = f_{\text{potential}} + f_{\text{drag}}$. In the overdamped limit, the inertial term is negligible, and the equation of motion becomes

[1] In liquid crystals, director rotation is coupled with hydrodynamic flow of the fluid, known as backflow. This coupling is modeled by Ericksen-Leslie theory, which is described in many textbooks on liquid crystals. In magnets, the magnetization precesses as well as relaxing, as modeled by the Landau-Lifshitz-Gilbert equation. For a historical perspective, see T. L. Gilbert, "Classics in Magnetics: A Phenomenological Theory of Damping in Ferromagnetic Materials," *IEEE Trans. Magn.* **40**, 3443 (2004).

J. V. Selinger, *Introduction to Topological Defects and Solitons*, Lecture Notes in Physics 1032, https://doi.org/10.1007/978-3-031-70200-6_4

$$0 = f_{\text{potential}} + f_{\text{drag}} = -\frac{\partial U}{\partial x} - \frac{\partial D}{\partial \dot{x}} = -\frac{\partial U}{\partial x} - b\dot{x} \quad \Rightarrow \quad b\dot{x} = -\frac{\partial U}{\partial x}. \tag{4.1}$$

Let us now apply the same method to the dynamics of the xy model. As a first step, we must decide whether to work with the director $\hat{\boldsymbol{n}}(\boldsymbol{r}, t)$ or the angle $\theta(\boldsymbol{r}, t)$. The disadvantage of working with the director is that we would need to impose a constraint to keep $\hat{\boldsymbol{n}}$ as a unit vector. Hence, we will work with θ instead. Next, we must determine the effective forces acting on θ. The effective potential is provided by the free energy of Eq. (3.1), and hence the force from the potential acting on θ is

$$f_{\text{potential}} = -\frac{\delta F}{\delta\theta(\boldsymbol{r}, t)} = K\nabla^2\theta. \tag{4.2}$$

Recall that K has units of (energy)(length) in 1D, energy in 2D, and energy/length in 3D, as discussed at the end of Sect. 3.1. The Rayleigh dissipation function is

$$D = \frac{1}{2}\gamma \int d\boldsymbol{r}|\dot{\boldsymbol{n}}|^2 = \frac{1}{2}\gamma \int d\boldsymbol{r}\dot{\theta}^2, \tag{4.3}$$

where γ is a rotational viscosity, and the integral is taken over all dimensions of the system. Hence, γ has units of (energy)(time)/length in 1D, (energy)(time)/length^2 in 2D, and (energy)(time)/length^3 in 3D. The force from drag acting on $\theta(\boldsymbol{r}, t)$ is

$$f_{\text{drag}} = -\frac{\delta F}{\delta\dot{\theta}(\boldsymbol{r}, t)} = -\gamma\dot{\theta}. \tag{4.4}$$

The equation of motion for $\theta(\boldsymbol{r}, t)$ then becomes

$$0 = f_{\text{potential}} + f_{\text{drag}} = K\nabla^2\theta - \gamma\dot{\theta} \quad \Rightarrow \quad \dot{\theta} = \left(\frac{K}{\gamma}\right)\nabla^2\theta, \tag{4.5}$$

which is equivalent to the diffusion equation. The ratio K/γ has units of length^2/time, like a diffusion constant, in all dimensions. We now want to see what this model for $\theta(\boldsymbol{r}, t)$ implies for the motion of defects and solitons.

4.1.1 Drag on Defects

First, let us calculate the drag force acting on a moving defect.[2] Suppose that the defect moves with velocity $\boldsymbol{u}$. Without loss of generality, we can choose our coordinate system so that the x-axis is aligned with the velocity vector $\boldsymbol{u} = u\hat{\boldsymbol{x}}$. We would like to determine how the Rayleigh dissipation function of Eq. (4.3) depends on u. For that calculation, we need the rotation rate $\dot{\theta}$ everywhere in the system. As

[2] To our knowledge, this calculation was originally done by H. Imura and K. Okano, "Friction Coefficient for a Moving Disinclination in a Nematic Liquid Crystal," *Phys. Lett. A* **42**, 403 (1973).

a simple approximation, which is reasonable when u is small, suppose that the angle field $\theta(\boldsymbol{r}, t)$ is still given by Eq. (3.7) but with a moving defect position, $x_d = ut$ and $y_d = 0$, so that

$$\theta(x, y, t) = q \tan^{-1}\left(\frac{y}{x - ut}\right) + \theta_0. \tag{4.6}$$

The rotation rate is then

$$\dot{\theta} = \frac{quy}{(x - ut)^2 + y^2}. \tag{4.7}$$

We convert this expression into polar coordinates centered on the defect, and substitute it back into Eq. (4.3), to obtain

$$D = \frac{1}{2}\gamma \int_0^\infty r dr \int_0^{2\pi} d\phi \left(\frac{qu \sin\phi}{r}\right)^2. \tag{4.8}$$

Like the free energy in Sect. 3.1, this dissipation function is logarithmically divergent at both small and large r. Once again, we handle these divergences by imposing a microscopic cutoff length r_{core} and long-distance cutoff length R. The integrated dissipation then becomes

$$D = \frac{1}{2}\pi\gamma q^2 u^2 \log\left(\frac{R}{r_{\text{core}}}\right). \tag{4.9}$$

The most important point about this result is that the dissipation is proportional to the velocity squared, analogous to a particle moving in a viscous fluid. Hence, when a defect moves with velocity $\boldsymbol{u}$, it experiences a drag force

$$\boldsymbol{f}_{\text{drag}} = -\frac{\partial D}{\partial \boldsymbol{u}} = -\left[\pi\gamma q^2 \log\left(\frac{R}{r_{\text{core}}}\right)\right]\boldsymbol{u}, \tag{4.10}$$

opposite to the velocity. For a point defect in a 2D system, this expression has units of force. For a line defect in a 3D system, the units are force per length of the defect. We can see that the drag coefficient is proportional to γ, the rotational viscosity of the ordered material. It is also proportional to q^2, where q is the topological charge of the defect.[3] Finally, it is proportional to $\log(R/r_{\text{core}})$, the ratio of the system size to the microscopic cutoff length. This size dependence occurs because

[3] In this simple theory, defects of topological charge $+q$ and $-q$ experience the same drag force. However, if one includes the coupling between director rotation and backflow in liquid crystals, then positive defects experience less drag than negative defects. This asymmetry between positive and negative defects has been seen in experiments, simulations, and analytic calculations, as discussed in X. Tang and J. V. Selinger, "Theory of Defect Motion in 2D Passive and Active Nematic Liquid Crystals," *Soft Matter* **15**, 587 (2019).

the defect involves a very spread-out distortion of the director field, as discussed in the previous chapters.

As an example of how to use the concept of drag on defects, consider the motion of two defects with opposite charges $\pm q$. Suppose the defects are separated by some distance $x_{12}(t) = x_2(t) - x_1(t)$ along the x-axis, as illustrated in Fig. 1.5b. The defects attract each other with a force

$$f_{\text{potential}} = -\frac{2\pi K q^2}{x_{12}}, \tag{4.11}$$

as derived in Eq. (3.16). Because of that attraction, they move toward each other. As they move, each defect experiences a drag force proportional to its velocity. Hence, the overdamped equations of motion are

$$\left.\begin{aligned} \pi\gamma q^2 \log(R/r_{\text{core}})\dot{x}_1 &= +2\pi K q^2/x_{12} \\ \pi\gamma q^2 \log(R/r_{\text{core}})\dot{x}_2 &= -2\pi K q^2/x_{12} \end{aligned}\right\} \Rightarrow \pi\gamma q^2 \log(R/r_{\text{core}})\frac{dx_{12}}{dt} = -\frac{4\pi K q^2}{x_{12}}. \tag{4.12}$$

Integrating the last equation gives

$$x_{12}(t) = \sqrt{\frac{8K(t_0 - t)}{\gamma \log(R/r_{\text{core}})}}. \tag{4.13}$$

This result shows that the defects come together and annihilate each other at the point of zero separation $x_{12} = 0$. The parameter t_0 is an integration constant, which represents the time at which the annihilation occurs. It is related to the initial separation by

$$t_0 = \frac{\gamma \log(R/r_{\text{core}})}{8K}\left(x_{12}^{\text{initial}}\right)^2. \tag{4.14}$$

That time is proportional to the rotational viscosity γ, because a higher viscosity makes the dynamics go more slowly. It is inversely proportional to the elastic constant K, because a higher elastic constant increases the attraction between opposite-charged defects.

This calculation provides a simple, approximate view of how opposite-charged defects come together and annihilate each other. It is not exact, because our treatment of the long-distance cutoff in Eq. (4.8) is not exact. We have assumed that a moving defect affects the rotation rate $\dot{\theta}$ all the way out to the full system size. More realistically, a moving defect affects $\dot{\theta}$ out to the distance of the next nearby defect. The integral should actually be cut off at a distance R given by the characteristic spacing between defects. For the problem of defect annihilation, R is of order x_{12}, which becomes smaller as the defects approach each other. Hence, the drag coefficient is not really constant. For a more precise theory, we should include the dependence of R on x_{12} in the differential equation (4.12).

Even if we include a cutoff distance that depends on x_{12}, the calculation still would not be exact. Rather, it would still be an approximation to say that the integral is cut off abruptly at a single distance. Actually, the dissipation integral (4.8) might depend on direction in the (x, y) plane, or on the system boundaries, or on the presence of any other defects in the system. These issues arise when we treat a moving defect, which is an extremely spread-out object, as if it were a localized particle.

In our view, we should just appreciate the particle-like theory in this section as a useful approximation. If we want a more precise prediction, we should go back to the partial differential equation (4.5) for $\theta(\boldsymbol{r}, t)$ everywhere in the system. Numerical solution of this partial differential equation can be difficult, because defects are positions where $\theta(\boldsymbol{r}, t)$ is undefined, and those positions are moving. In Chap. 5, we will show an alternative theoretical approach that deals with this difficulty.

4.1.2 Anomalous Drag on Defects

Some researchers[4] have criticized the theory in Sect. 4.1.1, because it assumes that the angle field $\theta(\boldsymbol{r}, t)$ is given by Eq. (3.7) but with a moving defect position. They argue that one should really solve the dynamic equation (4.5) with the constraint of a moving defect position. Their calculation shows that defect motion leads to director rotation out to a maximum length scale of $K/(\gamma u)$ away from the defect. Beyond that length, the director is approximately independent of time. Note that the length scale depends on the defect velocity—it diverges in the limit of slowly moving defects, but becomes smaller for faster motion. Based on this argument, the dissipation integral in Eq. (4.8) should be cut off at R or at $K/(\gamma u)$, whichever is smaller. If the system is truly infinite, then the second alternative is always smaller, and hence the integrated dissipation becomes

$$D = \frac{1}{2}\pi\gamma q^2 u^2 \log\left(\frac{K}{r_{\text{core}}\gamma u}\right). \tag{4.15}$$

The drag force is therefore

$$\boldsymbol{f}_{\text{drag}} = -\pi\gamma q^2 \boldsymbol{u} \log\left(\frac{K}{r_{\text{core}}\gamma u\sqrt{e}}\right). \tag{4.16}$$

[4] See G. Ryskin and M. Kremenetsky, "Drag Force on a Line Defect Moving through an Otherwise Undisturbed Field: Disclination Line in a Nematic Liquid Crystal," *Phys. Rev. Lett.* **67**, 1574 (1991); L. Radzihovsky, "Anomalous Energetics and Dynamics of Moving Vortices," *Phys. Rev. Lett.* **115**, 247801 (2015).

The important point about this result is that the drag force is not proportional to the velocity. Rather, it has an extra logarithmic factor. In that sense, the drag can be considered *anomalous*.

This argument is certainly correct for a single, isolated defect in an infinite system. However, it may be less relevant in a finite system (of size R), or in a system of many defects (with characteristic spacing R). In either of those cases, there is a crossover velocity $u_{\text{cross}} = K/(\gamma R)$ at which the cutoff length scale changes. For $u < u_{\text{cross}}$, a defect has conventional, linear drag given by Eq. (4.10). For $u > u_{\text{cross}}$, a defect has anomalous, logarithmic drag. It is not clear whether experiments would be able to detect the anomalous regime; that would depend on details of the experimental system.

4.1.3 Drag on Solitons

Let us now apply the same theoretical approach as in Sect. 4.1.1 to calculate the drag on a soliton moving with velocity $\boldsymbol{u}$. For low velocity, we assume that the angle field θ is still given by Eq. (3.20) but with a moving soliton position $x_0 = ut$, so that

$$\theta(x,t) = \pi + 4\tan^{-1}\left[\tanh\left(\frac{x-ut}{2\xi}\right)\right]. \tag{4.17}$$

Recall that ξ is the width of the soliton. The rotation rate then becomes

$$\dot{\theta} = -\frac{2u}{\xi}\text{sech}\left(\frac{x-ut}{2\xi}\right). \tag{4.18}$$

We put that expression into the dissipation of Eq. (4.3) and integrate over x to obtain

$$D = \frac{4\gamma u^2}{\xi}. \tag{4.19}$$

Hence, the drag force on a moving soliton is

$$\boldsymbol{f}_{\text{drag}} = -\frac{\partial D}{\partial \boldsymbol{u}} = -\left[\frac{8\gamma}{\xi}\right]\boldsymbol{u}. \tag{4.20}$$

For a point-like soliton in 1D, this expression has units of force. For a soliton line in 2D, the units are force per length of the soliton. For a soliton plane in 3D, the units are force per area of the soliton.

In this problem of a moving soliton, the dissipation integral does not have any issues of how to cut off a divergence, as we saw for a moving defect. Rather, the integral is nicely convergent, and the resulting drag coefficient is well-defined. This difference between solitons and defects occurs because a soliton involves only a short-range deformation of the director, over the length scale of ξ. Unlike a defect, a soliton does not have long-range effects on the director.

Overall, the dynamic theory of defects and solitons in Sect. 4.1 can be regarded as a type of *coarse-graining*.[5] In the context of sine-Gordon solitons, it has also been called the *collective coordinate method*. In this theoretical approach, we ignore all details about the director $\hat{\boldsymbol{n}}(\boldsymbol{r}, t)$ or the angle $\theta(\boldsymbol{r}, t)$ everywhere inside a material. Instead, we only keep a small number of particle-like degrees of freedom—just the positions of the defects. This coarse-graining is a powerful approach to simplify the description of complex systems.

4.2 Statistical Mechanics of Defects

Topological defects are important for statistical mechanics in two ways. First, as a very general phenomenon, the motion of defects is an essential part of how any system evolves from a disordered to an ordered state. Second, more specifically, the binding or unbinding of defects is essential for the Kosterlitz-Thouless transition of the 2D xy model in thermal equilibrium. Here, we briefly discuss each of these connections between defects and statistical mechanics.

4.2.1 Coarsening

For any statistical system, we can ask the question: How does the system evolve in time from a disordered state to an ordered state? What does the dynamic process look like?

As an example, consider a 2D xy model. At high temperature, the system is in a disordered state with high entropy, as illustrated in Fig. 4.1a. We can see that the blue arrows fluctuate randomly, independently of their neighbors. For a liquid crystal, these arrows represent local averages of molecular orientation, which are uncorrelated from one region to the next. For a magnet, they represent local averages of the magnetic dipole moment, uncorrelated from site to site. This state is so disordered that it cannot be well described in terms of defects, because there is no ordered region around any defect to draw a loop.

Now suppose that this system is suddenly cooled from high to low temperature. This sudden cooling is called a *quench*. Immediately after the quench, the system is still in the disordered state shown in Fig. 4.1a. However, the system now begins to develop orientational order, driven by the favorable energy of alignment. For a liquid crystal, the molecules align with their neighbors. For a magnet, the magnetic dipole moments align with nearby sites.

[5] For other aspects of the coarse-grained description of solitons, including interaction with changes in the background field, see C. Long and J. V. Selinger, "Coarse-Grained Theory for Motion of Solitons and Skyrmions in Liquid Crystals," *Soft Matter* **17**, 10437 (2021).

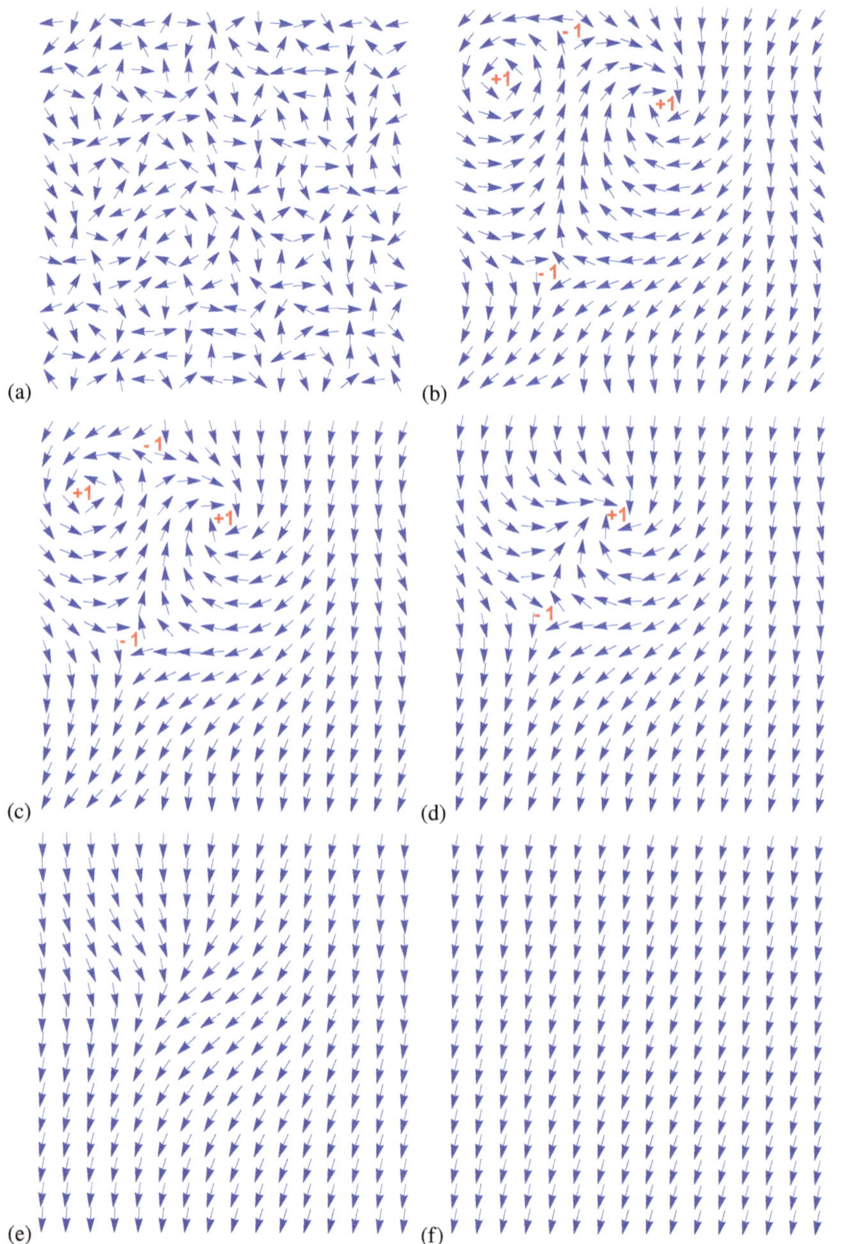

Fig. 4.1 Simulation of a 2D xy model, with periodic boundary conditions, which is quenched from a high-temperature disordered state down to low temperature. Part (**a**) shows the initial disordered state. In part (b), the system has rapidly evolved into a state with short-range orientational order, which has several defects of topological charge ± 1. In parts (**c**–**d**), opposite-charge defects slowly move together and annihilate each other. In parts (**e**–**f**), all of the defects have annihilated, and the system can evolve toward a uniform state

In different regions of the system, the orientational order begins to form in different directions. These regions are *not* separated by sharp domain walls, because such walls would cost too much free energy. Rather, the direction of orientational order must vary smoothly among these regions. This smooth variation often leads to topological defects. Hence, the system rapidly evolves into a state of short-range orientational order with many topological defects, as shown in Fig. 4.1b. The total topological charge of these defects is determined by the boundary conditions on the whole system. (For the example in Fig. 4.1, the system has periodic boundary conditions, and hence the total topological charge is zero.)

After the defects form, they interact through the free energy discussed in Sect. 3.2. Defects of opposite topological charge attract each other, and defects of like charge repel. Hence, opposite-charge defects slowly move toward each other and annihilate each other, through the dynamic mechanism discussed in Sect. 4.1.1. For example, one pair of opposite defects annihilates each other from Fig. 4.1c to d, and another pair annihilates each other from Fig. 4.1d to e. As the defects annihilate each other, the density of defects per area decreases, and the length scale of the orientational order increases. Eventually, after all of the defects are gone, the orientational order can become uniform, as shown in Fig. 4.1f.

The process of defect annihilation leading to longer-range order is called *coarsening*. It is a common phenomenon in many types of statistical systems. Apart from liquid crystals and magnets, it also occurs in particle physics and cosmology. The standard model of particle physics involves an abstract order parameter for the electroweak interaction. After the big bang, this order parameter is initially in a disordered state, but then it passes through a symmetry-breaking transition as the universe cools. In cosmological scenarios, this transition leads to topological defects in the order parameter, which are known as cosmic strings. They are analogous to defect lines in the 3D xy model, discussed in Sect. 1.3. The universe then coarsens through the annihilation of cosmic strings, following a process called the Kibble-Zurek mechanism. Researchers have considered liquid crystals as laboratory models for this cosmological mechanism.[6]

4.2.2 Kosterlitz-Thouless Transition

Apart from the general role of defects in coarsening, defects also have a very specific role in the phase transition between ordered and disordered states of the 2D xy model. This phase transition is known as the Kosterlitz-Thouless transition, and it is part of the research that was honored by the 2016 Nobel Prize in Physics.[7]

[6] See I. Chuang, R. Durrer, N. Turok, and B. Yurke, "Cosmology in the Laboratory: Defect Dynamics in Liquid Crystals," *Science* **251**, 1336 (1991).

[7] The Nobel Foundation provides excellent descriptions of the prize-winning research. See the website https://www.nobelprize.org/prizes/physics/2016/summary/, and click on Popular Science Background or Scientific Background.

We emphasize that this phase transition is different from the coarsening behavior described in Sect. 4.2.1. Coarsening is a dynamic process of approach toward thermal equilibrium, while the Kosterlitz-Thouless transition occurs in thermal equilibrium.

The simplest way to describe the Kosterlitz-Thouless transition is based on the competition between energy and entropy. To explain this competition, we must first review some fundamental statistical mechanics.

Background on Statistical Mechanics

In statistical mechanics, we work with the partition function Z, which is a sum over states

$$Z = \sum_{\text{states } i} e^{-E_i/k_B T}, \tag{4.21}$$

where i is a state of the system, E_i is the energy of that state, k_B is Boltzmann's constant, and T is the absolute temperature. The partition function is related to the free energy $\mathcal{F}$ by

$$\mathcal{F} = -k_B T \log Z, \tag{4.22}$$

where log means the natural logarithm (base e). Combining those two equations gives

$$e^{-\mathcal{F}/k_B T} = \sum_{\text{states } i} e^{-E_i/k_B T}. \tag{4.23}$$

Hence, the free energy $\mathcal{F}$ is a combination of the energies of all the states, combined through this summation. Note that $\mathcal{F}$ is *the* free energy of the whole system; it is not the free energy of any specific state, because it has been summed over all states.

In many physical problems, we want to classify the states. We can say that a system has many macrostates M, and each macrostate is composed of many microstates m. In that case, the summation breaks up into

$$e^{-\mathcal{F}/k_B T} = \sum_{\text{macrostates } M} \underbrace{\left[\sum_{\text{microstates } m} e^{-E_m/k_B T} \right]}_{e^{-F_M/k_B T}}. \tag{4.24}$$

Here, the inner sum is over all the microstates m that go into macrostate M, and the outer sum is over macrostates. We can write the inner sum as

(continued)

e^{-F_M/k_BT}, and hence define the free energy F_M of macrostate M. We can see that F_M is a coarse-grained quantity, which is intermediate between the energy E_m of a microstate and *the* free energy $\mathcal{F}$ of the whole system. It is not the energy of any specific microstate, because it has already been summed over microstates. It is not *the* free energy of the whole system, because it still needs to be summed over macrostates.

In general, we might classify states on many levels. We can group microstates into macrostates, and macrostates into super-macrostates, and super-macrostates into even-higher-macrostates, and so forth, until we reach the entire system. The *renormalization group* provides a systematic mathematical method to move up the hierarchy of states, gradually changing to a more coarse-grained description on longer length scales. It was recognized by the Nobel Prize in Physics to Kenneth Wilson in 1982. It is beyond the scope of this book.[8]

Let us now apply these general considerations of statistical mechanics to the 2D xy model. Here, each director configuration $\hat{\boldsymbol{n}}(\boldsymbol{r})$ is a macrostate, which includes many possible microstates. The microstates are all of the microscopic configurations of atoms and molecules that correspond to a single director configuration $\hat{\boldsymbol{n}}(\boldsymbol{r})$. We never explicitly think about these microstates, but we have already implicitly summed over them to get the macrostate free energy $F = \frac{1}{2}K\int d\boldsymbol{r}|\nabla\hat{\boldsymbol{n}}|^2$ of Eq. (3.1).

In *most* of this book, the macrostate free energy is exactly what we need, because it tells us the properties of a specific director configuration. It allows us to determine the optimal director configuration by minimizing the free energy. For example, it shows us that the free energy of a configuration with a defect is $F_{\text{defect}} = \pi K q^2 \log(R/r_{\text{core}})$ in Eq. (3.10).

This Sect. 4.2.2 is an exception. Here, we do not want to know the properties of a single director configuration. Instead, we want to group all the macrostates with one defect *at a specific position* into a single super-macrostate with one defect *at any position*. As a simple model, let us consider only defects of topological charge $q = \pm 1$, and neglect the higher charges. In that case, each macrostate has the same free energy $F_{\text{defect}} = \pi K \log(R/r_{\text{core}})$. The number of macrostates in the group is just the number of possible positions to put a defect, $N_{\text{positions}}$. Hence, Eq. (4.24) gives

$$e^{-F_{\text{defect at any position}}/k_BT} = N_{\text{positions}} e^{-F_{\text{defect}}/k_BT}. \tag{4.25}$$

[8] For an introduction to the renormalization group, see S.-k. Ma, *Modern Theory of Critical Phenomena* (Routledge, 1976).

Taking the logarithm of both sides of that equation gives

$$F_{\text{defect at any position}} = F_{\text{defect}} - k_B T \log N_{\text{positions}} = F_{\text{defect}} - T S_{\text{positions}}, \tag{4.26}$$

where $S_{\text{positions}} = k_B \log N_{\text{positions}}$ is the positional entropy associated with the range of possible positions for the defect. We can estimate that $N_{\text{positions}}$ is the system area πR^2 divided by the defect core area πr_{core}^2, so that $S_{\text{positions}} = 2k_B \log(R/r_{\text{core}})$. Here, we note that F_{defect} and $S_{\text{positions}}$ depend on the ratio R/r_{core} in exactly the same way. Hence, we can combine them to obtain

$$F_{\text{defect at any position}} = (\pi K - 2k_B T) \log\left(\frac{R}{r_{\text{core}}}\right). \tag{4.27}$$

By comparison, the free energy for no defect is $F_{\text{no defect}} = 0$.

The important point about this result is that $\Delta F = F_{\text{defect at any position}} - F_{\text{no defect}}$ changes sign at the critical temperature

$$T_{KT} = \frac{\pi K}{2k_B}. \tag{4.28}$$

For $T > T_{KT}$, the free energy ΔF is *negative*, which means that the entropic benefit of forming an isolated defect exceeds the energetic cost. In that case, many isolated defects will form throughout the system. By comparison, for $T < T_{KT}$, the free energy ΔF is *positive*, and hence isolated defects are suppressed. However, the free energy cost of a pair of oppositely charged defects (calculated in Sect. 3.2) scales only as $\log(r_{12}/r_{\text{core}})$, which is much less than the entropic benefit proportional to $\log(R/r_{\text{core}})$. Hence, many of these neutral pairs will form.

For this reason, the Kosterlitz-Thouless transition at T_{KT} is a *defect binding-unbinding* transition. Above the transition, defects are unbound and strongly affect the director field. Below the transition, defects are bound in neutral pairs.

Kosterlitz and Thouless developed a more rigorous version of this argument, using the renormalization group. Their theory provides a foundation for models of defect-mediated transitions in many experimental systems, including 2D superfluid films and melting of 2D solids (which will be discussed in Sect. 9.4.2).

Prequel to Defects: Variable Magnitude of Order

5

In Hollywood, a prequel means a movie that is made *after* the original movie, but tells a story that occurs *before* the original movie. The most famous example is *Star Wars* episodes 1–3, which were produced many years after *Star Wars* episodes 4–6, but show events occurring earlier. These movies provide background information, so that viewers can understand how Darth Vader became so evil.

This chapter does the same thing with topological defects. We now shift to a theoretical approach that is more fundamental that the theory presented in the previous chapters, and we show that this approach provides background information about defects. Based on this approach, we offer some comments on the relationship between topology and physics.

5.1 Theory with Variable Magnitude

For a more fundamental theory, we consider the concept of an order parameter with variable magnitude. In Sect. 1.1, we began our discussion of the xy model with a vector order parameter $\boldsymbol{M}(\boldsymbol{r})$. For a magnet, this order parameter is the average magnetic moment at position $\boldsymbol{r}$. It can be written in terms of Cartesian components as $\boldsymbol{M} = (M_x, M_y)$, or in terms of magnitude and direction as $\boldsymbol{M} = M\hat{\boldsymbol{n}} = (M\cos\theta, M\sin\theta)$. In Chaps. 1–4, we assumed that the magnitude $M = |\boldsymbol{M}|$ is constant, so that we would only need to think about the unit vector $\hat{\boldsymbol{n}}$ or angle θ. Now, we relax that assumption, and allow both the magnitude and direction of $\boldsymbol{M}$ to vary.

Because the magnitude of order can vary, we must consider how the free energy depends on the magnitude, even in a uniform system with no spatial gradients. Researchers generally assume the form

$$F_{\text{bulk}} = \int d\boldsymbol{r} \left[-\frac{1}{2}a|\boldsymbol{M}|^2 + \frac{1}{4}b|\boldsymbol{M}|^4 + \ldots \right], \tag{5.1}$$

J. V. Selinger, *Introduction to Topological Defects and Solitons*, Lecture Notes in Physics 1032, https://doi.org/10.1007/978-3-031-70200-6_5

relative to the disordered state ($F = 0$ at $\boldsymbol{M} = 0$). Here, $a > 0$ and $b > 0$ are phenomenological coefficients. Why this form? There are two motivations:[1]

1. One motivation is based on Landau theory for second-order phase transitions, which depends purely on symmetry and smoothness of functions. The free energy must be a scalar, which can be expanded as a power series about $\boldsymbol{M} = 0$. This series can only include the even powers $|\boldsymbol{M}|^2$, $|\boldsymbol{M}|^4$, ... The odd powers $|\boldsymbol{M}|$, $|\boldsymbol{M}|^3$, ... are not allowed because they are not smooth functions of $\boldsymbol{M}$ around $\boldsymbol{M} = 0$. Hence, we can express the free energy as the series in Eq. (5.1). The series is conventionally truncated after the fourth-order term, because $|\boldsymbol{M}|$ is small near a second-order phase transition. The same truncation is often used away from a second-order transition, just because it provides a simple model for the free energy, even if it is not mathematically justified.
2. A second motivation comes from mean-field theory for a lattice of magnetic dipoles $\boldsymbol{m}_i$ in the xy plane, interacting through the Hamiltonian $H = -J \sum \boldsymbol{m}_i \cdot \boldsymbol{m}_j$, with the expectation value $\boldsymbol{M} = \langle \boldsymbol{m}_i \rangle$. The mean-field probability distribution function for each dipole is $\rho(\boldsymbol{m}_i)$, and the mean-field free energy is $F = \langle H \rangle + k_B T \langle \rho \log \rho \rangle$. Like Landau theory, the mean-field free energy can be expanded as a power series for small M, as in Eq. (5.1). Unlike Landau theory, the mean-field free energy diverges as $|\log(M_{\max} - M)|$ when M approaches the maximum saturated value $M_{\max}$. Researchers often use the power series and neglect the logarithmic divergence, just for simplicity.

Regardless of the motivation, let us use Eq. (5.1) to represent the free energy as a function of $\boldsymbol{M} = (M_x, M_y)$. A plot of this function is shown in Fig. 5.1. Its shape is commonly described as a sombrero (Mexican hat), or as the bottom of a wine bottle. The most important feature of this free energy is that it has a continuous set of ground states, which occur at $\boldsymbol{M} = (M_0 \cos\theta, M_0 \sin\theta)$, with

$$M_0 = \sqrt{\frac{a}{b}} \tag{5.2}$$

and any angle θ. This set of ground states is shown by the thick blue circle in the figure. Physically, all of these states have the same magnitude of order M_0, but have order in different directions. All of these states have a free energy of

$$F_{\text{bulk}} = \int d\boldsymbol{r} \left[-\frac{a^2}{b} \right]. \tag{5.3}$$

[1] In our previous book J. V. Selinger, *Introduction to the Theory of Soft Matter: From Ideal Gases to Liquid Crystals* (Springer, 2016), see Chapter 4 for a discussion of Landau theory. See Sections 2.2–2.3 for mean-field theory of the Ising model, and 10.4 for Maier-Saupe mean-field theory of nematic liquid crystals.

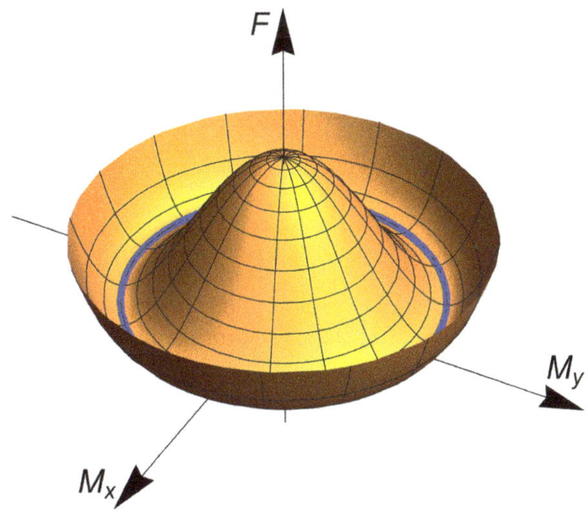

Fig. 5.1 Free energy of a uniform system, given in Eq. (5.1), as a function of order parameter $\boldsymbol{M} = (M_x, M_y)$. The thick blue line shows the continuous set of ground states

In field theory, the term *Goldstone mode* describes variation about the blue circle, varying θ at constant $M = M_0$ and at constant free energy.

In addition to the free energy associated with uniform $\boldsymbol{M}$, we also need the free energy associated with variations of $\boldsymbol{M}(\boldsymbol{r})$ as a function of position $\boldsymbol{r}$. The simplest model assumes that this free energy density is proportional to the square of the spatial gradient $\nabla\boldsymbol{M}$. Hence, the total gradient free energy is the integral

$$\begin{aligned} F_{\text{gradient}} &= \frac{1}{2}L\int d\boldsymbol{r}|\nabla\boldsymbol{M}|^2 = \frac{1}{2}L\int d\boldsymbol{r}(\partial_i M_j)(\partial_i M_j) \\ &= \frac{1}{2}L\int d\boldsymbol{r}\left[|\nabla M_x|^2 + |\nabla M_y|^2\right], \end{aligned} \tag{5.4}$$

where L is an elastic coefficient. If we write $\boldsymbol{M} = (M\cos\theta, M\sin\theta)$ in terms of its magnitude and direction, then this expression simplifies to

$$F_{\text{gradient}} = \frac{1}{2}L\int d\boldsymbol{r}\left[|\nabla M|^2 + M^2|\nabla\theta|^2\right]. \tag{5.5}$$

We can now combine Eqs. (5.1) and (5.5) to obtain

$$\begin{aligned} F &= \int d\boldsymbol{r}\left[-\frac{1}{2}a|\boldsymbol{M}|^2 + \frac{1}{4}b|\boldsymbol{M}|^4 + \frac{1}{2}L|\nabla\boldsymbol{M}|^2\right] \\ &= \int d\boldsymbol{r}\left[-\frac{1}{2}aM^2 + \frac{1}{4}bM^4 + \frac{1}{2}L|\nabla M|^2 + \frac{1}{2}LM^2|\nabla\theta|^2\right]. \end{aligned} \tag{5.6}$$

This is our general expression for the free energy in the variable-magnitude theory.

As one special case of this theory, suppose that the magnitude of the order is constant, so that $M = M_0$ everywhere in the system. In that case, Eq. (5.6) simplifies to $F = \text{constant} + \frac{1}{2} L M_0^2 \int d\boldsymbol{r} |\nabla\theta|^2$. This expression is equivalent to the free energy of Eq. (3.1) in the unit-vector theory, provided that the elastic coefficients are related by

$$K = L M_0^2. \tag{5.7}$$

We can now use the variable-magnitude theory to study topological defects. We would like to calculate the optimum configurations of *both* $\theta(\boldsymbol{r})$ and $M(\boldsymbol{r})$ around a defect. The Euler-Lagrange equations are

$$\frac{\delta F}{\delta M(\boldsymbol{r})} = -aM + bM^3 - L\nabla^2 M + LM|\nabla\theta|^2 = 0, \tag{5.8a}$$

$$\frac{\delta F}{\delta \theta(\boldsymbol{r})} = -LM^2\nabla^2\theta - 2LM(\nabla M)\cdot(\nabla\theta) = 0. \tag{5.8b}$$

To describe a defect of topological charge q, we can use the solution for θ that we already found in Eq. (3.5), which is

$$\theta(r,\phi) = q\phi + \theta_0, \quad \nabla\theta = \frac{q}{r}\hat{\boldsymbol{\phi}}, \quad |\nabla\theta|^2 = \frac{q^2}{r^2}, \tag{5.9}$$

in polar coordinates (r, ϕ). In that case, M must be a function of r alone, independent of ϕ, satisfying

$$-aM + bM^3 - L\frac{d^2M}{dr^2} - \frac{L}{r}\frac{dM}{dr} + \frac{Lq^2}{r^2}M = 0. \tag{5.10}$$

To solve this equation, we rescale M by $M_0 = \sqrt{a/b}$ defined above, so that $\tilde{M} = M/M_0$. We also rescale r by

$$\rho_{\text{core}} = \sqrt{\frac{L}{a}}, \tag{5.11}$$

so that $\tilde{r} = r/\rho_{\text{core}}$. We will see later that ρ_{core} is related to the defect core radius. The scaled differential equation then becomes

$$-\tilde{M} + \tilde{M}^3 - \frac{d^2\tilde{M}}{d\tilde{r}^2} - \frac{1}{\tilde{r}}\frac{d\tilde{M}}{d\tilde{r}} + \frac{q^2}{\tilde{r}^2}\tilde{M} = 0. \tag{5.12}$$

We do not know an exact solution to this differential equation. A numerical solution for defects of topological charge $q = \pm 1$ is plotted in Fig. 5.2. It can be approximated reasonably well by

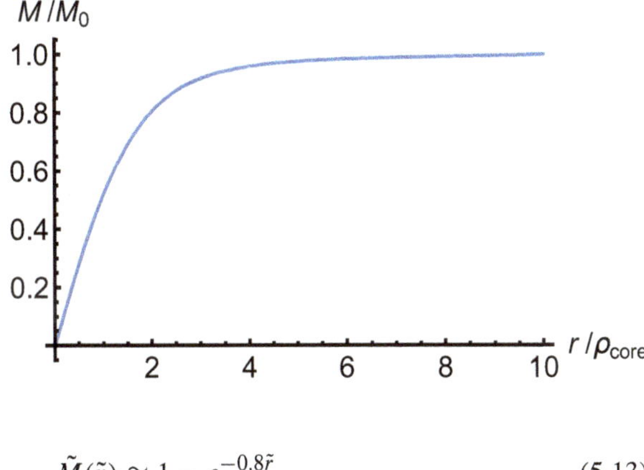

Fig. 5.2 Numerical solution of Eq. (5.12), with $q = \pm 1$, showing how the magnitude of the order parameter varies with the distance from the center of a defect

$$\tilde{M}(\tilde{r}) \approx 1 - e^{-0.8\tilde{r}}. \tag{5.13}$$

In the original, unscaled variables, the corresponding approximation is

$$M(r) \approx M_0(1 - e^{-0.8r/\rho_{\text{core}}}). \tag{5.14}$$

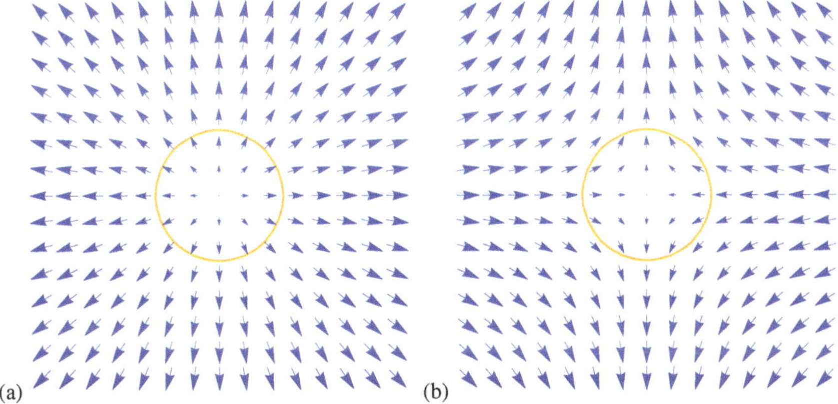

Fig. 5.3 Visualization of topological defects in the variable-magnitude theory. The blue arrows represent the vector field $\boldsymbol{M}(\boldsymbol{r})$, with both magnitude and direction. The gold circles show the radius ρ_{core} around each defect. (**a**) $q = +1$. (**b**) $q = -1$

The important physical point of these calculations is that the magnitude of order varies around a defect. Right at the center of the defect, the magnitude is exactly zero. Away from the center, the magnitude increases approximately linearly with r, out to a distance of order ρ_{core}. Beyond that distance, the magnitude asymptotically approaches the bulk value M_0. Figure 5.3a and b shows the vector fields $\boldsymbol{M}(\boldsymbol{r})$ around defects of topological charge $q = +1$ and $q = -1$, respectively. In these visualizations, the length of the arrows represents the magnitude $M(\boldsymbol{r})$, and the direction of the arrows represents the orientation. The gold circles are drawn with the radius ρ_{core}. Far from the centers, the magnitude is approximately constant, and these figures look just like Fig. 1.4a and e. However, close to the centers, Fig. 5.3a and b provides a more microscopic view of what happens in the defect cores. Note that $\boldsymbol{M}(\boldsymbol{r})$ is a smooth function everywhere—including at the centers, where it goes smoothly to zero.

Using this solution for $\boldsymbol{M}(\boldsymbol{r})$, we can calculate the free energy of a defect of topological charge $q = \pm 1$ in the variable-magnitude theory. For this calculation, we put the solutions of Eqs. (5.9) and (5.14) into the free energy of Eq. (5.6). We subtract off the free energy of a uniform, defect-free state given in Eq. (5.3) to get the difference ΔF. We then integrate over the entire system. Unlike our previous free energy calculation in Sect. 3.1, this new calculation does *not* have an ultraviolet divergence; the integral is well-behaved in the limit of $r \to 0$. Hence, we do not need to impose a nonzero lower limit on the integral. The variable-magnitude theory has solved that problem by providing a model for the behavior inside the defect core. However, we *still* have an infrared divergence in the limit of $r \to \infty$; the variable-magnitude theory cannot solve that problem. Hence, we must again impose an upper limit on the integral, based on the system size R. The integral then becomes

$$\Delta F \approx \pi \underbrace{L \underbrace{\left(\frac{a}{b}\right)}_{M_0^2}}_{K} \left[0.4 + \log\left(R \underbrace{\sqrt{\frac{a}{L}}}_{\frac{1}{\rho_{\text{core}}}}\right)\right] \approx \pi K \log\left(\frac{R}{0.7\rho_{\text{core}}}\right), \tag{5.15}$$

in the limit of $R \gg \rho_{\text{core}} = \sqrt{L/a}$. By comparison, our previous calculation in the unit-vector theory gave $\Delta F = \pi K \log(R/r_{\text{core}})$ for a defect with $q = \pm 1$, in Eq. (3.10). These results are consistent provided that the lower cutoff r_{core} is taken to be $0.7\rho_{\text{core}}$. The factor of 0.7 is not very important, because the variable-magnitude theory has a gradual change in $M(r)$ around ρ_{core}, not a sharp change at any specific radius.

In addition to the free energy, we can also use the variable-magnitude theory to calculate the drag acting on a defect. For that calculation, we need to repeat the procedure in Sect. 4.1.1, but now using the full order parameter $\boldsymbol{M}$ instead of unit vector $\hat{\boldsymbol{n}}$ or angle θ. As a first step, we note that any time-dependent changes in the magnitude or direction of $\boldsymbol{M}$ must lead to energy dissipation. This dissipation is represented by the Rayleigh dissipation function D. The simplest model for that function is

$$D = \frac{1}{2}\Gamma \int d\boldsymbol{r} |\dot{\boldsymbol{M}}|^2, \tag{5.16}$$

where Γ is a viscosity in the variable-magnitude theory. If we write $\boldsymbol{M} = (M\cos\theta, M\sin\theta)$ in terms of its magnitude and direction, then this expression simplifies to

$$D = \frac{1}{2}\Gamma \int d\boldsymbol{r} \left[\dot{M}^2 + M^2\dot{\theta}^2\right]. \tag{5.17}$$

If the magnitude of the order is constant, so that $M = M_0$ at all times, then Eq. (5.17) simplifies to $D = \frac{1}{2}\Gamma M_0^2 \int d\boldsymbol{r}\dot{\theta}^2$. This expression is equivalent to Eq. (4.3), provided that the viscosities are related by

$$\gamma = \Gamma M_0^2. \tag{5.18}$$

This relationship between viscosities Γ and γ is analogous to the relationship between elastic constants L and K in Eq. (5.7).

We can now calculate the dissipation associated with a moving defect. Following the procedure in Sect. 4.1.1, we assume that a defect is moving with velocity vector $\boldsymbol{u} = u\hat{\boldsymbol{x}}$. We suppose that the order parameter around the moving defect is

$$\boldsymbol{M}(x, y, t) = \boldsymbol{M}_{\text{defect}}(x - ut, y) = M(x - ut, y)(\cos\theta(x - ut, y), \sin\theta(x - ut, y)), \tag{5.19}$$

where $\theta(x - ut, y) = q\tan^{-1}(y/(x - ut)) + \theta_0$, and $M(x - ut, y)$ is the magnitude calculated approximately in Eq. (5.14), with $q = \pm 1$. We differentiate with respect to time to get $\dot{\boldsymbol{M}}$, put that result into Eq. (5.16), and integrate over the entire system. The integral does not have an ultraviolet divergence for short distances, but it has an infrared divergence for long distances, and hence we must cut it off at the system size R. In the end, we obtain

$$D = \frac{1}{2}\pi \underbrace{\Gamma M_0^2}_{\gamma} u^2 \log\left(\frac{R}{1.1\rho_{\text{core}}}\right), \tag{5.20}$$

in the limit of $R \gg \rho_{\text{core}} = \sqrt{L/a}$. This result is equivalent to Eq. (4.9) from the unit-vector theory, with topological charge $q = \pm 1$, provided that the lower cutoff r_{core} is taken to be $1.1\rho_{\text{core}}$. The factor of 1.1 is not important, because the variable-magnitude theory does not have any sharp change at a core radius.

From these results, we see that the drag force acting on a moving defect (with $q = \pm 1$) is $f_{\text{drag}} = -\partial D/\partial u = -[\pi\gamma\log(R/r_{\text{core}})]u$. Hence, the dynamic predictions of the variable-magnitude theory are basically the same as the predictions of the unit-vector theory.

5.2 Comparison of Unit-Vector and Variable-Magnitude Theories

Now that we have presented the variable-magnitude theory, we can discuss how it is related to the unit-vector theory, and how researchers can choose which theory to use.

The variable-magnitude theory is more fundamental, because it involves more degrees of freedom. It considers possible variations in the magnitude as well as the direction of $\boldsymbol{M}(\boldsymbol{r})$. The unit-vector theory is a higher-level, more coarse-grained theory, because it eliminates the magnitude and considers only the direction of $\boldsymbol{M}(\boldsymbol{r})$. The unit-vector theory can be regarded as a limiting case of the variable-magnitude theory, in the limit where the magnitude is strongly constrained to be close to M_0, i.e. near the blue circle in Fig. 5.1.

The main advantage of the variable-magnitude theory is that it is *more specific* and hence provides *more detailed predictions*. It makes a specific model for F_{bulk} as a function of the magnitude M, and hence predicts the detailed structure of the defect cores shown in Fig. 5.3. It also shows how the defect free energy and drag force depend on the parameters in F_{bulk}. All of this information is unavailable in the unit-vector theory.

By contrast, the main advantage of the unit-vector theory is that it is *more general*, as it does not depend on any specific model for F_{bulk}. For example, in some system, F_{bulk} might not be well-described by the power series in Eq. (5.1)—it might have a logarithmic divergence as M approaches the maximum saturated value. In that case, all the calculations in the variable-magnitude theory would need to be redone, but the unit-vector theory would be unchanged.

Apart from those considerations, some readers might ask which theory is *simpler* for calculations. The answer depends on the type of calculations.

For analytic, paper-and-pencil calculations, the unit-vector theory is usually simpler. It is normally more convenient to work with fewer variables, rather than more variables. Furthermore, the partial differential equation for the angle $\theta(\boldsymbol{r})$ is Laplace's equation, which is linear. It can often be solved exactly in specific geometries, using standard mathematical methods like conformal mapping, image constructions, or series expansions. By contrast, the partial differential equation for the magnitude $M(\boldsymbol{r})$ is nonlinear, and hence more difficult to solve.

For numerical calculations, the variable-magnitude theory is often simpler. The variable-magnitude $\boldsymbol{M}(\boldsymbol{r})$ is an ordinary vector field, which is well-defined everywhere, even at defects. This type of vector field is well-suited for numerical algorithms. In comparison, $\hat{\boldsymbol{n}}(\boldsymbol{r})$ is more difficult for numerical calculations, because it has singularities at defects, and because of the need to maintain the constraint $|\hat{\boldsymbol{n}}| = 1$. Likewise, $\theta(\boldsymbol{r})$ is also difficult for numerical calculations, because it also has singularities at defects, and because it is not a single-valued function of position.[2]

[2] Whenever we go around a defect, θ changes by $360° = 2\pi$ radians. Hence, at any specific point, it is impossible to say whether $\theta = 1°$ or $361°$ or $721°$ or .... This issue can create major problems for numerical calculations. For example, suppose the software shows that two nearby points have $\theta = 1°$ and $359°$. Naively, if we subtract those angles, we might think that the derivative of θ is huge. However, the physical difference between the orientations is very small. To avoid that problem, we should really calculate derivatives of $\cos\theta$ and $\sin\theta$ (or equivalently $\hat{\boldsymbol{n}}$ or $\boldsymbol{M}$), rather than derivatives of θ itself.

5.3 Comparison of Topology and Physics

Apart from the comparison between theories, we can also make some general comments about the relationship between topology and physics. These comments are our own personal perspective, which may not be shared by all researchers.

In general, topology makes a sharp distinction between what is *allowed* and what is *forbidden*. For our example of the xy model, it is *allowed* to rotate the orientation of $\hat{\boldsymbol{n}}$ continuously inside a system. It is *forbidden* to change the magnitude of $\hat{\boldsymbol{n}}$, or to introduce discontinuities in $\hat{\boldsymbol{n}}$, or to change $\hat{\boldsymbol{n}}$ on a boundary.

By contrast, in physics, all transformations are possible, but they cost different amounts of free energy. There is a low free energy cost for rotating the orientation of $\boldsymbol{M}$ continuously inside a system. There is a high free energy cost for changing the magnitude of $\boldsymbol{M}$, or for introducing discontinuities, or for changing the orientation on a boundary.

From this point of view, topology is an approximation to physics. In this approximation, we say that certain transformations are *allowed* because they have a low free energy cost, and other transformations are *forbidden* because they have a high free energy cost. This approximation works extremely well if there is a wide separation of free energy scales, so that the forbidden transformations cost much more free energy than the allowed transformations. The approximation works less well if the free energy scales are not so different.

For a specific example of the relationship between topology and physics, we can re-examine the 1D xy model configurations of Fig. 1.1. In Chap. 1, we argued that it is impossible to transform any configuration with winding number $q = 0$ into a configuration with $q = 1$. That argument is based on topology—we cannot go from $q = 0$ to $q = 1$ without making a forbidden change in $\hat{\boldsymbol{n}}(x)$. Now, let us relax that constraint. Suppose that the arrows represent the order parameter $\boldsymbol{M}(x)$. We *can* change the magnitude of $\boldsymbol{M}(x)$, although that change might require a high free energy. In that case, it *is* possible to transform a configuration with $q = 0$ into a configuration with $q = 1$. Indeed, Fig. 5.4 shows a specific example of this transformation, from $q = 0$ in row a to $q = 1$ in row g. The essential point is that the magnitude of $\boldsymbol{M}(x)$ passes through zero in the center of row d. Hence, row d must have a higher free energy than rows a or g. To change the winding number, we need to get over this free energy barrier.

The question is now: How high is the free energy barrier? To answer this question, we can use the variable-magnitude theory of Sect. 5.1. At the center of row d is a region where $\boldsymbol{M}(x)$ is suppressed. Outside this region, $\boldsymbol{M}(x)$ has the bulk magnitude $M_0 = \sqrt{a/b}$; inside the region, $\boldsymbol{M}(x)$ goes to smoothly to zero. This region is similar to the core of a topological defect, and it has the same optimal size $\rho_{\text{core}} = \sqrt{L/a}$ found in Eq. (5.11). The free energy cost of this region is different from the free energy of a topological defect in Eq. (5.15) because it is a 1D rather than a 2D integral, so it becomes

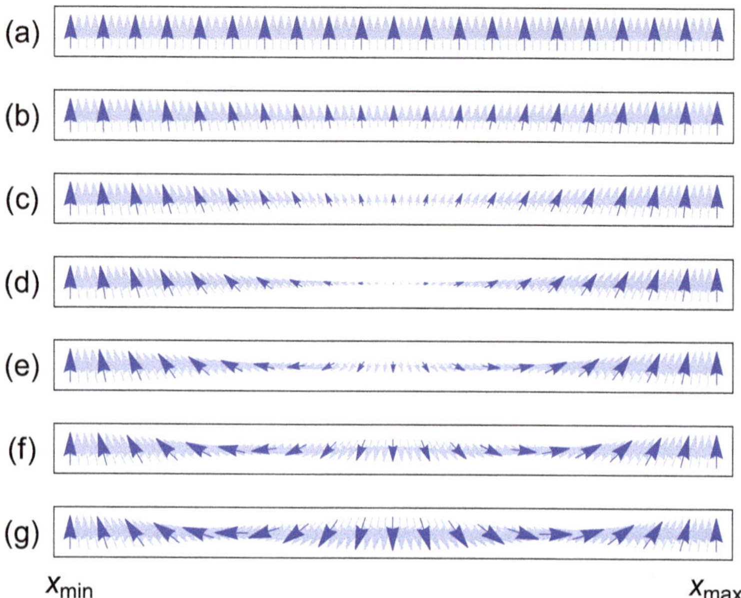

Fig. 5.4 Schematic illustration of how a 1D xy model *might* change its winding number q if the order parameter $\boldsymbol{M}(x)$ can pass through zero. Row a has $q = 0$, and row g has $q = 1$. This process does *not* normally occur in liquid crystals or magnets, because the free energy barrier associated with row d is too high. This figure should be compared with Fig. 1.1, which shows transformations that do not change the winding number, and do not require going over the high free energy barrier

$$\Delta F \approx \frac{L^{1/2}a^{3/2}}{b} = \frac{LM_0^2}{\rho_{\text{core}}} = \frac{K}{\rho_{\text{core}}}. \tag{5.21}$$

Hence, we can distinguish between two cases:

- Case 1: Suppose the energetic parameters a and b are very high (with a specific ratio between them, so that $M_0 = \sqrt{a/b}$ is fixed). The length scale $\rho_{\text{core}} = \sqrt{L/a}$ is small, in comparison with other lengths in the system. The free energy barrier $\Delta F = K/\rho_{\text{core}}$ is high, in comparison with other energies in the system. The change in winding number shown in Fig. 5.4 is practically impossible, because it requires too much free energy. In this case, topology is an excellent approximation to physics.
- Case 2: Suppose the energetic parameters a and b are not so high (and they still have the same ratio, so that $M_0 = \sqrt{a/b}$ is fixed). The length scale $\rho_{\text{core}} = \sqrt{L/a}$ is similar to other lengths in the system, and the free energy barrier $\Delta F = K/\rho_{\text{core}}$ is similar to other energies in the system. The change in winding number shown in Fig. 5.4 does not require too much free energy, and hence it might actually occur. In this case, topology is less useful as an approximation to physics.

The point of this book is that many physical systems really are in case 1, so that core radii are small (in comparison with other length scales), and free energy barriers are high (in comparison with other energy scales). In this case, the topological arguments in the other chapters are very useful for describing what physical processes can occur, without requiring unrealistic energies.

Further Issues: Defect Phase/Orientation, Charge Density, Curvature

6

In this chapter, we consider a few further issues: the definition of defect phase or orientation, the concept of topological charge density, and the interaction of defects with curved environments. These topics are more advanced than the previous chapters, and some people may choose to skip them on a first reading. They are sometimes useful in understanding the physics of defects in the xy model, and they have some generalizations in Part II of the book.

6.1 Defect Phase or Orientation

In Sect. 3.1, when we calculated the director configuration around a defect that minimizes the free energy, we found $\hat{\boldsymbol{n}} = (\cos\theta, \sin\theta)$ with

$$\theta(r, \phi) = q\phi + \theta_0, \tag{6.1}$$

in polar coordinates (r, ϕ), or

$$\theta(x, y) = q \tan^{-1}\left(\frac{y}{x}\right) + \theta_0. \tag{6.2}$$

in Cartesian coordinates (x, y). We have extensively discussed the meaning of the coefficient q, which is the winding number or topological charge of the defect. However, we have not yet considered the meaning of the additive constant θ_0. Let us now consider the physical significance of that parameter.

In Sects. 6.1.1 and 6.1.2, we interpret θ_0 in two distinct physical cases: superfluid vs. liquid crystal or magnet. In Sect. 6.1.3, we show how θ_0 affects the interaction between defects in *both* of these physical cases.

J. V. Selinger, *Introduction to Topological Defects and Solitons*, Lecture Notes in Physics 1032, https://doi.org/10.1007/978-3-031-70200-6_6

6.1.1 Superfluid

As a first model, let us consider a physical system like a superfluid, where the xy order is a quantum phase. In this situation, the unit vector $\hat{\boldsymbol{n}}(\boldsymbol{r})$ does not point in any direction in real space; rather, it has some direction in an abstract order parameter space. It is often written as a complex number $\psi(\boldsymbol{r}) = e^{i\theta(\boldsymbol{r})}$. The unit vector and the complex number are two equivalent ways of describing the same mathematical object.

In the complex number notation, the order around a defect can be written as

$$\psi(r, \phi) = e^{iq\phi} e^{i\theta_0}. \tag{6.3}$$

From that expression, we can see directly that the parameter θ_0 just shifts the quantum phase of ϕ by the same amount everywhere around the defect. Hence, θ_0 can be regarded as the phase of the defect. As usual in quantum mechanics, a constant phase is not directly observable; only phase differences are observable.

6.1.2 Liquid Crystal or Magnet

As a second model, suppose we have a liquid crystal or a magnet. In this case, the unit vector $\hat{\boldsymbol{n}}(\boldsymbol{r})$ actually points in some direction in real space. It might point radially outward from or inward toward the defect, or tangentially around the defect, or in some other direction.

Suppose we want to characterize where $\hat{\boldsymbol{n}}(\boldsymbol{r})$ points radially outward from the defect.[1] This alignment occurs wherever the director angle θ and the polar coordinate ϕ represent the same orientation in the plane, so that

$$\theta = \phi \pmod{2\pi}. \tag{6.4}$$

Here, the notation (mod 2π) means that $\theta = \phi + 2\pi k$, where k is an integer. Any two angles that differ by 0, $\pm 2\pi$, $\pm 4\pi$, ..., represent the same physical orientation. We can now combine Eqs. (6.1) and (6.4) to obtain

$$q\phi + \theta_0 = \phi \pmod{2\pi}. \tag{6.5}$$

Hence, the special directions where $\hat{\boldsymbol{n}}(\boldsymbol{r})$ points radially outward occur at $\theta = \phi = \psi$, where

[1] This issue was first considered by A. J. Vromans and L. Giomi, "Orientational Properties of Nematic Disclinations," *Soft Matter* **12**, 6490 (2016), in the context of 2D nematic defects, which we will discuss in Chap. 7. The description was later generalized by X. Tang and J. V. Selinger, "Orientation of Topological Defects in 2D Nematic Liquid Crystals," *Soft Matter* **13**, 5481 (2017), to cover other models, including the 2D xy model discussed here.

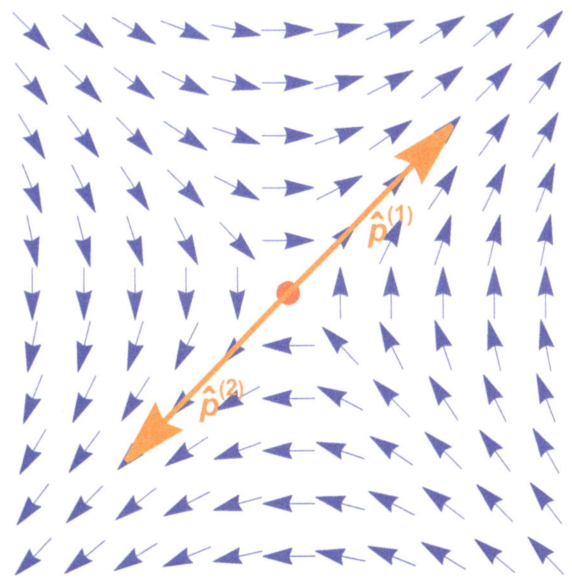

Fig. 6.1 Defect of topological charge $q = -1$ in an xy model. The defect structure has twofold rotational symmetry, with orientation characterized by the vectors $\hat{\boldsymbol{p}}^{(1,2)} = (\cos\psi^{(1,2)}, \sin\psi^{(1,2)})$, shown in orange. This example has $\theta_0 = \pi/2$, and hence $\psi^{(1)} = \pi/4$, $\psi^{(2)} = 5\pi/4$

$$\psi = \frac{\theta_0}{1-q} \quad \left(\bmod \frac{2\pi}{|1-q|}\right). \tag{6.6}$$

This orientation can be represented as a unit vector $\hat{\boldsymbol{p}} = (\cos\psi, \sin\psi)$.

As a specific example, suppose we have a defect of topological charge $q = -1$, as shown in Fig. 6.1. Equation (6.6) then reduces to $\psi = \frac{1}{2}\theta_0 \pmod{\pi}$. It has two distinct solutions, $\psi^{(1)} = \frac{1}{2}\theta_0$ and $\psi^{(2)} = \frac{1}{2}\theta_0 + \pi$, where the orientational order is pointing outward. These two solutions correspond to two distinct unit vectors $\hat{\boldsymbol{p}}^{(1)} = (\cos(\frac{1}{2}\theta_0), \sin(\frac{1}{2}\theta_0))$ and $\hat{\boldsymbol{p}}^{(2)} = (\cos(\frac{1}{2}\theta_0 + \pi), \sin(\frac{1}{2}\theta_0 + \pi)) = -\hat{\boldsymbol{p}}^{(1)}$, represented by the orange arrows in Fig. 6.1. Hence, the defect structure has twofold rotational symmetry. The orientation of the twofold symmetric defect is described by the angle $\psi^{(1)}$ or $\psi^{(2)}$, with the understanding that these two angles represent the same orientation. Equivalently, the orientation is described by the unit vector $\hat{\boldsymbol{p}}^{(1)}$ or $\hat{\boldsymbol{p}}^{(2)}$, with the understanding that these two unit vectors represent the same orientation. If we want to represent the defect orientation by a single-valued mathematical object, we can construct a tensor of rank two,

$$\boldsymbol{T} = \frac{1}{2}\left(\hat{\boldsymbol{p}}^{(1)}\hat{\boldsymbol{p}}^{(1)} + \hat{\boldsymbol{p}}^{(2)}\hat{\boldsymbol{p}}^{(2)}\right), \tag{6.7}$$

using the tensor or outer product. The orientation is then described by the eigenvector of that tensor, corresponding to eigenvalue 1.

The same argument applies to topological defects with any topological charge $q \neq +1$, so that we are not dividing by zero in Eq. (6.6). The equation then has

$|1-q|$ distinct solutions for ψ, where the orientational order is pointing outward. These solutions correspond to $|1-q|$ distinct unit vectors $\hat{\boldsymbol{p}}$. Hence, the defect structure has $|1-q|$-fold rotational symmetry. Its orientation can be described by any of these ψ angles, or any of these $\hat{\boldsymbol{p}}$ vectors, with the understanding that all of these angles and all of these unit vectors represent the same orientation. If we want to represent the defect orientation by a single-valued mathematical object, we can construct a tensor of rank $|1-q|$,

$$\boldsymbol{T} = \underbrace{\underbrace{\hat{\boldsymbol{p}}^{(1)}\hat{\boldsymbol{p}}^{(1)}\cdots\hat{\boldsymbol{p}}^{(1)}}_{|1-q|\text{ factors}} + \underbrace{\hat{\boldsymbol{p}}^{(2)}\hat{\boldsymbol{p}}^{(2)}\cdots\hat{\boldsymbol{p}}^{(2)}}_{|1-q|\text{ factors}} + \cdots + \underbrace{\hat{\boldsymbol{p}}^{(|1-q|)}\hat{\boldsymbol{p}}^{(|1-q|)}\cdots\hat{\boldsymbol{p}}^{(|1-q|)}}_{|1-q|\text{ factors}}}_{|1-q|\text{ terms}}. \tag{6.8}$$

The orientation is then described by the generalized eigenvectors of that tensor.[2]

The topological charge of $q=+1$ is a special case, which must be considered separately. If $q=+1$, then the parameter θ_0 does *not* determine a defect orientation ψ. Rather, θ_0 determines the structure of the defect. For $\theta_0=0$, the orientational order points radially outward everywhere, as shown in Fig. 1.4a. For $\theta_0=\pi$, the orientational order points radially inward everywhere, as in Fig. 1.4b. For $\theta_0=\pi/2$ or $-\pi/2$, the orientational order points tangentially counter-clockwise or clockwise, as in Fig. 1.4c,d. If θ_0 is an intermediate angle, then the orientational order has an intermediate direction, between radial and tangential. For all of these cases, the defect has perfect rotational symmetry, and hence it does not have any orientation in the plane.

Summarizing the results of Sects. 6.1.1 and 6.1.2, we have:

- If $\hat{\boldsymbol{n}}(\boldsymbol{r})$ points in some direction in real space (as in a liquid crystal or magnet), and the topological charge is $q \neq +1$, then the defect phase parameter θ_0 determines the defect orientation ψ through Eq. (6.6).
- If $\hat{\boldsymbol{n}}(\boldsymbol{r})$ points in some direction in real space (as in a liquid crystal or magnet), and the topological charge is $q=+1$, then the defect phase parameter θ_0 determines the defect structure (radially outward or inward, tangential counter-clockwise or clockwise, or intermediate).
- If $\hat{\boldsymbol{n}}(\boldsymbol{r})$ points in some direction in an abstract order parameter space (as in a superfluid), then the phase parameter θ_0 is just a phase, nothing more.

[2] The generalized eigenvectors can be defined by the following construction: Consider a test vector $\hat{\boldsymbol{b}}=(\cos\beta,\sin\beta)$, calculate the scalar $f(\beta)=\boldsymbol{T}\cdot\hat{\boldsymbol{b}}\cdot\hat{\boldsymbol{b}}\cdots\hat{\boldsymbol{b}}$, and find the maxima of $f(\beta)$. These maxima occur at the angles $\beta=\psi\ (\mathrm{mod}\ 2\pi/|1-q|)$, and the corresponding $\hat{\boldsymbol{b}}$ vectors are generalized eigenvectors.

6.1.3 Effect of Phase on Interaction Between Defects

Let us now apply the considerations of defect phase to the interaction between two defects. The argument in this section applies to a liquid crystal or magnet, and also to a superfluid.

In Sect. 3.2, we calculated the director configuration that minimizes the free energy around two defects of charges q_1 and q_2 at positions $\boldsymbol{r}_1 = (x_1, y_1)$ and $\boldsymbol{r}_2 = (x_2, y_2)$. We found $\hat{\boldsymbol{n}} = (\cos\theta, \sin\theta)$ with

$$\theta(x, y) = q_1 \tan^{-1}\left(\frac{y - y_1}{x - x_1}\right) + q_2 \tan^{-1}\left(\frac{y - y_2}{x - x_2}\right) + \theta_0. \tag{6.9}$$

In this expression, there is only one phase parameter θ_0, which affects the phases of *both* defects. There is no way to specify the phase of each individual defect separately. From that perspective, the expression appears to be missing one degree of freedom.

The physical reason for the missing degree of freedom is that we have minimized the free energy subject only the constraints that there is a defect of charge q_1 at position $\boldsymbol{r}_1$, and another defect of charge q_2 at position $\boldsymbol{r}_2$. We did not specify anything about the phases of these defects. It turns out that there is a family of solutions, in which we vary θ_0 and change the phases of both defects together. All of these solutions have the same free energy. However, if we tried to vary the relative phase of the defects, then the free energy would change. One specific relative phase gives the lowest free energy, and we have already found it; it is included implicitly in Eq. (6.9).

As an example, Fig. 6.2a shows the director field of Eq. (6.9) with one defect of $q_1 = -1$ at $\boldsymbol{r}_1 = (0, 0)$, and another defect of $q_2 = +1$ at $\boldsymbol{r}_2 = (1, 0)$. The overall phase parameter is $\theta_0 = 0$. The -1 defect has a phase $\theta_1 = \pi$, as we can see because:

- It is oriented vertically, so that $\hat{\boldsymbol{p}} = \hat{\boldsymbol{y}}$ and $\psi = \frac{1}{2}\theta_1 = \pi/2$.
- Immediately to the right of the defect, the director is $\hat{\boldsymbol{n}} = -\hat{\boldsymbol{x}}$ with $\theta = \pi$.

The $+1$ defect has a phase $\theta_2 = 0$, as we can see because:

- The director points outward everywhere around the defect.
- Immediately to the right of the defect, the director is $\hat{\boldsymbol{n}} = +\hat{\boldsymbol{x}}$ with $\theta = 0$.

The relative phase $\theta_1 - \theta_2 = \pi$ is the optimum relative phase for these two topological charges in these two positions. We can change the overall phase without increasing the free energy, but we cannot change the relative phase without increasing the free energy.

If we want to change the relative phase, we must go back to the Euler-Lagrange equation (3.2), and solve it in a domain that excludes the defect cores, with boundary conditions that specify the phase of each individual defect. This calculation can be

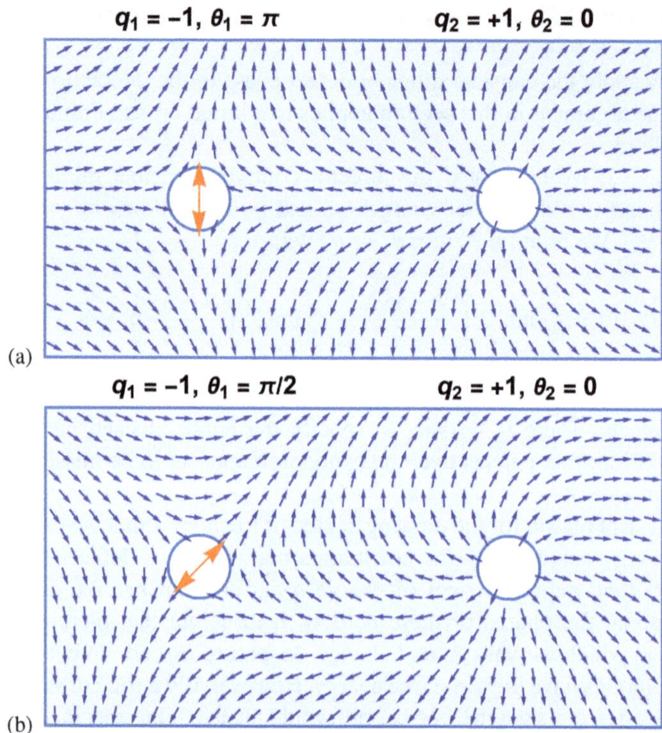

Fig. 6.2 Director field between two defects with topological charges -1 and $+1$. (**a**) Optimal relative phase $\delta\theta = 0$. (**b**) Non-optimal relative phase $\delta\theta = \pi/2$. The orange arrows represent the orientation of the -1 defect. The light blue region represents the domain for $\theta(x, y)$, in which we have solved the Euler-Lagrange equation

done by the mathematical method of conformal mapping.[3] The solution is

$$\theta(x, y) = q_1 \tan^{-1}\left(\frac{y - y_1}{x - x_1}\right) + q_2 \tan^{-1}\left(\frac{y - y_2}{x - x_2}\right) + \frac{\delta\theta}{2}\left[1 + \frac{\log(|\boldsymbol{r} - \boldsymbol{r}_1|^2) - \log(|\boldsymbol{r} - \boldsymbol{r}_2|^2)}{\log(|\boldsymbol{r}_1 - \boldsymbol{r}_2|^2) - \log(r_{\text{core}}^2)}\right] + \Theta, \tag{6.10}$$

in the limit of small core radius r_{core}. The parameters $\delta\theta$ and Θ are integration constants, which must be chosen to match the two defect phases θ_1 and θ_2. Those

[3] X. Tang and J. V. Selinger, "Orientation of Topological Defects in 2D Nematic Liquid Crystals," *Soft Matter* **13**, 5481 (2017).

conditions give[4]

$$\Theta = \theta_1 - q_2 \tan^{-1}\left(\frac{y_1 - y_2}{x_1 - x_2}\right), \tag{6.11}$$

$$\delta\theta = \theta_2 - \theta_1 - \underbrace{\left[q_1 \tan^{-1}\left(\frac{y_2 - y_1}{x_2 - x_1}\right) - q_2 \tan^{-1}\left(\frac{y_1 - y_2}{x_1 - x_2}\right)\right]}_{\text{optimal value of } \theta_2 - \theta_1}.$$

Note that $\delta\theta$ is *not* defined as $\theta_2 - \theta_1$; rather, it is the difference between the actual and optimal values of $\theta_2 - \theta_1$. The example in Fig. 6.2a has $\delta\theta = 0$, which is the optimal relative phase.

For comparison, Fig. 6.2b shows an example with non-optimal relative phase. Here, we still have one defect of $q_1 = -1$ at $\boldsymbol{r}_1 = (0, 0)$, and another defect of $q_2 = +1$ at $\boldsymbol{r}_2 = (1, 0)$. The -1 defect has now been rotated so that its phase is $\theta_1 = \pi/2$ (and its defect orientation is $\psi = \frac{1}{2}\theta_1 = \pi/4$). The $+1$ defect still has the director field pointing radially outward, so that its phase is $\theta_2 = 0$. This pair of defects has $\delta\theta = \pi/2$. We can see that this non-optimal relative phase creates extra distortion in the director field throughout the system, from the logarithmic terms in Eq. (6.10). This extra distortion must cost extra free energy.

To determine the free energy cost of the extra distortion, we put the solution for $\theta(x, y)$ from Eq. (6.10) into the free energy integral of Eq. (3.1). We integrate over the entire system, following the same procedure as in Sect. 3.2. The result is

$$F_{\text{int}} = 2\pi K q_1 q_2 \log\left(\frac{L_x}{|\boldsymbol{r}_1 - \boldsymbol{r}_2|}\right) + \frac{\pi K (\delta\theta)^2}{2} \frac{\log(|\boldsymbol{r}_1 - \boldsymbol{r}_2|/(2r_{\text{core}}))}{[\log(|\boldsymbol{r}_1 - \boldsymbol{r}_2|/r_{\text{core}})]^2}, \tag{6.12}$$

where L_x is the system size. Here, the first term is the Coulomb-like logarithmic interaction between the defects, as in Eq. (3.15), which is repulsive for like charges and attractive for opposite charges. The second term is a new contribution, which favors phase alignment between the defects with the relative phase $\delta\theta = 0$. It creates a long-range aligning torque of

$$-\frac{\partial F}{\partial(\delta\theta)} = -\pi K \delta\theta \frac{\log(|\boldsymbol{r}_1 - \boldsymbol{r}_2|/(2r_{\text{core}}))}{[\log(|\boldsymbol{r}_1 - \boldsymbol{r}_2|/r_{\text{core}})]^2}, \tag{6.13}$$

which decays very slowly for $|\boldsymbol{r}_1 - \boldsymbol{r}_2| \gg r_{\text{core}}$.

[4] In these expressions, $\tan^{-1}(\Delta y/\Delta x)$ should be understood as $\arctan(\Delta x, \Delta y)$. In many computer languages, the arctan function with two arguments is chosen so that it matches the correct quadrant for both Δx and Δy, and it ranges from $-\pi$ to π. For example, $\arctan(1, -1) = -\pi/4$ and $\arctan(-1, 1) = 3\pi/4$. In general, $\arctan(\Delta x, \Delta y)$ and $\arctan(-\Delta x, -\Delta y)$ differ by π.

6.2 Topological Charge Density

In the previous chapters, we saw that the topological charge q is somewhat analogous to an electric charge. For example, it leads to the interaction between defects that resembles Coulomb's law in 2D. This analogy leads to the following question: In electrostatics, it is often useful to think about an electric charge density, which can be integrated to give the total electric charge. Is it useful, or even possible, to define a similar topological charge density?

As we understand it, the concept of topological charge density is not really a concept of topology, because it is a local property. Topology is concerned with global properties of a structure, such as the winding number. Topological charge density is more of a physics concept, applied to topology.

In this section, we consider how topological charge density can be defined—first a longer discussion about defects, and then a very brief discussion about solitons. The argument here applies specifically to the xy model, but related considerations will occur for other models in Part II.

6.2.1 Topological Charge Density of Defects

In the 2D xy model, the total topological charge q_{enclosed} inside a region R is defined as a 1D line integral along a loop around that region R

$$q_{\text{enclosed in } R} = \frac{1}{2\pi}\oint_{\text{loop around } R} d\theta = \frac{1}{2\pi}\oint_{\text{loop around } R} (\nabla\theta)\cdot d\boldsymbol{l}. \tag{6.14}$$

We can transform this integral by using Stokes' theorem, which tells us that the line integral of any vector field $\boldsymbol{v}(\boldsymbol{r})$ along a boundary is the area integral of $\hat{z}\cdot\nabla\times\boldsymbol{v}$ in the interior

$$\oint_{\text{loop around } R} \boldsymbol{v}\cdot d\boldsymbol{l} = \int_{\text{region } R} d\boldsymbol{r}\left(\hat{z}\cdot\nabla\times\boldsymbol{v}\right) = \int_{\text{region } R} d\boldsymbol{r}\left(\partial_x v_y - \partial_y v_x\right). \tag{6.15}$$

Here, we are interested in the vector field $\boldsymbol{v} = \nabla\theta$, and hence Stokes' theorem gives

$$\begin{aligned} q_{\text{enclosed in } R} &= \int_{\text{region } R} d\boldsymbol{r}\sigma(\boldsymbol{r}) = \frac{1}{2\pi}\int_{\text{region } R} d\boldsymbol{r}\left(\hat{z}\cdot\nabla\times(\nabla\theta)\right) \\ &= \frac{1}{2\pi}\int_{\text{region } R} d\boldsymbol{r}\left[\nabla\cdot(\nabla\theta\times\hat{z})\right] = \frac{1}{2\pi}\int_{\text{region } R} d\boldsymbol{r}\left(\partial_x\partial_y\theta - \partial_y\partial_x\theta\right). \end{aligned} \tag{6.16}$$

This expression looks like the integral of a topological charge density $\sigma(\boldsymbol{r})$ over the region R, with

$$\sigma(\boldsymbol{r}) = \frac{1}{2\pi}\hat{\boldsymbol{z}} \cdot \nabla \times (\nabla\theta) = \frac{1}{2\pi}\nabla \cdot \left(\nabla\theta \times \hat{\boldsymbol{z}}\right) = \frac{1}{2\pi}\left(\partial_x \partial_y \theta - \partial_y \partial_x \theta\right). \tag{6.17}$$

Note that $\sigma(\boldsymbol{r})$ has dimensions of 1/length^2. We know that the integral in Eq. (6.16) is zero for any region that *does not* include a defect, and it is finite for any region that *does* include a defect. That mathematical property is exactly the definition of a Dirac delta function. Hence, we can conclude that the topological charge density is

$$\sigma(\boldsymbol{r}) = q_{\text{defect}}\delta(\boldsymbol{r} - \boldsymbol{r}_{\text{defect}}). \tag{6.18}$$

This equation tells us that the topological charge density has an infinite spike at any defect, and it is zero everywhere else. In that respect, it is analogous to the electric charge density around a point particle, like an electron. If the system has more than one defect, then the topological charge density is a sum of spikes at each defect, just as the electric charge density is a sum of spikes at each point charge.

Some readers might be uncomfortable with this expression for the topological charge density in terms of derivatives of $\theta(\boldsymbol{r})$, because of the usual mathematical statement that partial derivatives commute (or equivalently, that the curl of a gradient is zero). However, these usual statements are only true when the function is sufficiently well-behaved. Here, $\theta(\boldsymbol{r})$ is sufficiently well-behaved everywhere *except* at a defect, and indeed the difference of partial derivatives is a Dirac delta function, which is zero everywhere except at a defect. A defect is a singular point where $\theta(\boldsymbol{r})$ is not sufficiently well-behaved, and hence its partial derivatives do not commute (and the curl of the gradient is not zero).

We can also express the topological charge in terms of the director field $\hat{\boldsymbol{n}}(\boldsymbol{r})$ rather than the angle field $\theta(\boldsymbol{r})$. Because $\hat{\boldsymbol{n}} = (\cos\theta, \sin\theta)$, we have $\partial_k\theta = \epsilon_{3ij} n_i \partial_k n_j$, and hence

$$\begin{aligned}\sigma(\boldsymbol{r}) &= \frac{1}{2\pi}\epsilon_{3lk}\partial_l(\partial_k\theta) = \frac{1}{2\pi}\epsilon_{3lk}\partial_l(\epsilon_{3ij} n_i \partial_k n_j) = \frac{1}{2\pi}(\delta_{li}\delta_{kj} - \delta_{lj}\delta_{ki})\partial_l(n_i \partial_k n_j) \\ &= \frac{1}{2\pi}\nabla \cdot \left[\hat{\boldsymbol{n}}(\nabla \cdot \hat{\boldsymbol{n}}) - (\hat{\boldsymbol{n}} \cdot \nabla)\hat{\boldsymbol{n}}\right]. \end{aligned} \tag{6.19}$$

That expression for the topological charge density is interesting because it leads us to an analogue of the electric field. In a 2D universe, the differential form of Gauss's law is

$$\sigma(\boldsymbol{r}) = \frac{1}{2\pi}\nabla \cdot \boldsymbol{E}. \tag{6.20}$$

Hence, for a topological defect in the xy model, the analogue of an electric field is the vector field inside the divergence in Eq. (6.17) or (6.19), which is

$$\boldsymbol{E}_{\text{eff}} = \nabla\theta \times \hat{\boldsymbol{z}} = (\partial_y\theta, -\partial_x\theta) = \hat{\boldsymbol{n}}(\nabla \cdot \hat{\boldsymbol{n}}) - (\hat{\boldsymbol{n}} \cdot \nabla)\hat{\boldsymbol{n}}. \tag{6.21}$$

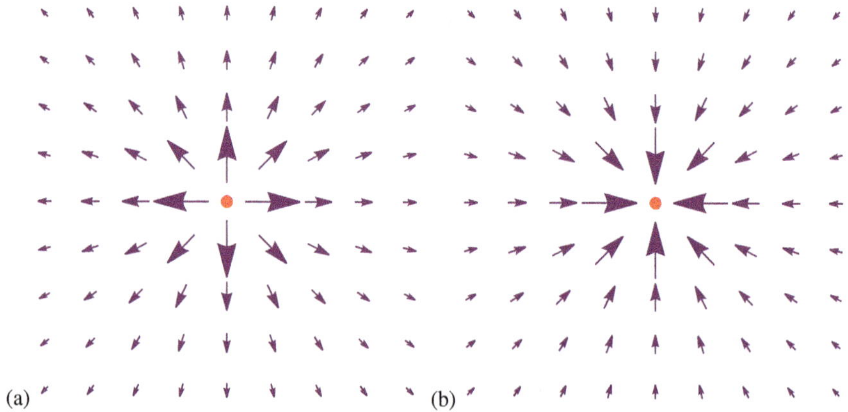

Fig. 6.3 Visualization of the vector field $\boldsymbol{E}_{\text{eff}}(\boldsymbol{r})$, defined in Eq. (6.21), around a defect. (**a**) Topological charge $q > 0$. (**b**) Topological charge $q < 0$. The vector field $\boldsymbol{E}_{\text{eff}}(\boldsymbol{r})$, not the director $\hat{\boldsymbol{n}}(\boldsymbol{r})$, is analogous to the electric field in Gauss's law

This vector field points radially outward from all defects with topological charge $q > 0$, and radially inward to all defects with topological charge $q < 0$, as shown in Fig. 6.3. In the context of liquid crystals, it is regarded as the splay vector plus the bend vector.[5]

Some researchers also define the analogue of an electrostatic potential, which is a scalar function $\Phi(\boldsymbol{r})$ that satisfies

$$\boldsymbol{E}_{\text{eff}} = -\nabla\Phi. \tag{6.22}$$

The study of this type of function is part of the branch of mathematics known as *potential theory*. The potential $\Phi(\boldsymbol{r})$ is related to the angle $\theta(\boldsymbol{r})$ by

$$\nabla\theta = \nabla\Phi \times \hat{z}, \quad \partial_x\theta = \partial_y\Phi, \quad \partial_y\theta = -\partial_x\Phi. \tag{6.23}$$

If $\Phi(\boldsymbol{r})$ is a sufficiently well-behaved function, so that its partial derivatives commute, then the Euler-Lagrange equation (3.2) for $\theta(\boldsymbol{r})$ is automatically satisfied. Equations (6.17–6.18) for $\theta(\boldsymbol{r})$ then become

$$\nabla^2\Phi = -2\pi\sigma(\boldsymbol{r}) = -2\pi q_{\text{defect}}\delta(\boldsymbol{r} - \boldsymbol{r}_{\text{defect}}) \tag{6.24}$$

for $\Phi(\boldsymbol{r})$, analogous to Poisson's equation.

[5] The splay scalar is $\nabla \cdot \hat{\boldsymbol{n}}$, and hence the splay vector is $\hat{\boldsymbol{n}}(\nabla \cdot \hat{\boldsymbol{n}})$. The bend vector is $-(\hat{\boldsymbol{n}} \cdot \nabla)\hat{\boldsymbol{n}}$. Because $\hat{\boldsymbol{n}}$ is a unit vector, the bend vector can also be written as $\hat{\boldsymbol{n}} \times (\nabla \times \hat{\boldsymbol{n}})$, which may be more recognizable to some readers.

These mathematical analogies are occasionally useful to determine $\theta(\boldsymbol{r})$ in complex 2D configurations with defects. The general strategy is:

1. Solve Poisson's equation (6.24) for $\Phi(\boldsymbol{r})$, perhaps using methods familiar from electrostatics.
2. Calculate $\nabla\theta = \nabla\Phi \times \hat{z}$.
3. Integrate to find $\theta(\boldsymbol{r})$.

As a simple example, suppose we have a single defect of topological charge q at the origin, in an infinite system. In polar coordinates (r, ϕ), the potential function is $\Phi = -q \log r +$ constant. Its gradient is $\nabla\Phi = -(q/r)\hat{\boldsymbol{r}}$, and hence $\nabla\theta = (q/r)\hat{\boldsymbol{\phi}}$. Integrating that gradient gives $\theta = q\phi + \theta_0$, which we already know.

So far, we have defined the topological charge density in terms of the angle $\theta(\boldsymbol{r})$ or the unit vector $\hat{\boldsymbol{n}}(\boldsymbol{r})$. We can actually generalize this definition to the variable-magnitude theory of Chap. 5.[6] In Eq. (6.19), let us replace $\hat{\boldsymbol{n}}(\boldsymbol{r})$ by the variable-magnitude $\boldsymbol{M}(\boldsymbol{r})/M_0$. These vector fields are identical far from a defect, but they have different magnitudes near a defect. The alternative definition of topological charge density then becomes[7]

$$\sigma(\boldsymbol{r}) = \frac{1}{2\pi M_0^2}\nabla \cdot [\boldsymbol{M}(\nabla \cdot \boldsymbol{M}) - (\boldsymbol{M} \cdot \nabla)\boldsymbol{M}] . \tag{6.25}$$

If we write $\boldsymbol{M} = (M\cos\theta, M\sin\theta)$ in terms of its magnitude and direction, then this expression becomes

$$\sigma(\boldsymbol{r}) = \frac{M}{\pi M_0^2}\hat{z} \cdot (\nabla M) \times (\nabla\theta). \tag{6.26}$$

Around a defect of topological charge q, we have $\nabla\theta = (q/r)\hat{\boldsymbol{\phi}}$ in polar coordinates (r, ϕ), and hence the topological charge density simplifies further to

$$\sigma(\boldsymbol{r}) = \frac{qM}{\pi r M_0^2}\frac{\partial M}{\partial r}. \tag{6.27}$$

The integral of this topological charge density is

[6] This analysis is a simplified version of C. D. Schimming and J. Viñals, "Singularity Identification for the Characterization of Topology, Geometry, and Motion of Nematic Disclination Lines," *Soft Matter* **18**, 2234 (2022). That article developed the theory for 3D nematic order, as we will discuss in Sect. 8.5. Here, we apply it to the xy model.

[7] Note that $-(\boldsymbol{M} \cdot \nabla)\boldsymbol{M}$ is not necessarily equal to $\boldsymbol{M} \times (\nabla \times \boldsymbol{M})$, because $\boldsymbol{M}$ does not have constant magnitude.

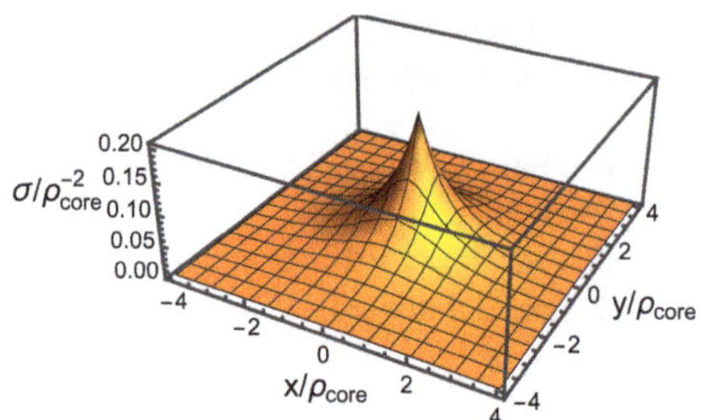

Fig. 6.4 Smeared-out version of the topological charge density $\sigma(\boldsymbol{r})$ in the variable-magnitude theory, from Eq. (6.29)

$$\int d\boldsymbol{r}\sigma(\boldsymbol{r}) = 2\pi \int_0^\infty r dr \frac{qM}{\pi r M_0^2}\frac{\partial M}{\partial r} = \frac{q}{M_0^2}\left[M(\infty)^2 - M(0)^2\right] = q, \tag{6.28}$$

which is the total topological charge.

In the variable-magnitude theory, the topological charge density is *not* a Dirac delta function spike at the defect; rather, it is spread out around the defect. For example, with our approximate form $M(r) \approx (1 - e^{-0.8r/\rho_{\rm core}})$ from Eq. (5.14), it becomes

$$\sigma(\boldsymbol{r}) = \frac{0.8q}{\pi r \rho_{\rm core} M_0^2} e^{-0.8r/\rho_{\rm core}}(1 - e^{-0.8r/\rho_{\rm core}}), \tag{6.29}$$

which is plotted in Fig. 6.4. Hence, the variable-magnitude theory smears the topological charge density over a region of size $\rho_{\rm core}$, and it keeps the appropriate integral.

Overall, the concept of topological charge density works well for defects in the xy model. However, we should warn readers: There is no guarantee that this concept will work for other types of defects in Part II. We will need to consider that issue on a case-by-case basis. The principles of topology apply to global properties of structures, and they might or might not be expressible in terms of local properties, such as charge density.

6.2.2 Topological Charge Density of Solitons

We can briefly apply the concept of a topological charge density to solitons, rather than defects. In the 1D xy model, the winding number of a soliton is defined as

$$q = \frac{1}{2\pi}\int d\theta = \frac{1}{2\pi}\int \left(\frac{d\theta}{dx}\right) dx = \frac{1}{2\pi}\int \hat{z}\cdot\left(\hat{\boldsymbol{n}} \times \frac{d\hat{\boldsymbol{n}}}{dx}\right) dx. \tag{6.30}$$

Here, we have again used the relation $\partial_k\theta = \epsilon_{3ij} n_i \partial_k n_j$ because $\hat{\boldsymbol{n}} = (\cos\theta, \sin\theta)$. To define a topological charge density, we would like to say that q is an integral of the charge density $\lambda(x)$,

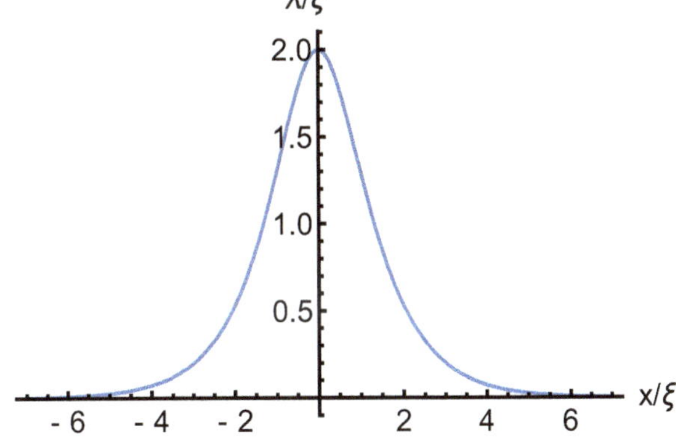

Fig. 6.5 Plot of the topological charge density $\lambda(x)$ for a single soliton centered at $x_0 = 0$, from Eq. (6.33). This example corresponds to the $\hat{\boldsymbol{n}}(x)$ plots in Fig. 2.1 and the $\theta(x)$ plot in Fig. 3.3

$$q = \frac{1}{2\pi} \int \lambda(x) dx. \tag{6.31}$$

Because the integral is 1D, $\lambda(x)$ is a *linear* charge density, with dimensions of 1/length. Fortunately, Eq. (6.30) is already in the correct form; we do not need to transform it using Stokes' theorem. Hence, we can define $\lambda(x)$ as

$$\lambda(x) = \frac{d\theta}{dx} = \hat{z} \cdot \left(\hat{\boldsymbol{n}} \times \frac{d\hat{\boldsymbol{n}}}{dx} \right). \tag{6.32}$$

For example, with our solution $\theta(x) = \pi + 4 \tan^{-1}[\tanh(x - x_0)/(2\xi)]$ from Eq. (3.20), the charge density becomes

$$\lambda(x) = \frac{2}{\xi} \operatorname{sech} \left(\frac{x - x_0}{\xi} \right). \tag{6.33}$$

This function is plotted in Fig. 6.5, and we can see that it is a peak with width of order ξ.

We have already used the topological charge density $\lambda(x)$ in Figs. 2.1–2.5, where it provides the yellow shading to help visualize solitons. Apart from computer graphics, we cannot think of any other uses for this concept. Even so, the concept is available for researchers who might develop ways to use it.

6.3 Defects in Curved Environments

Another important physical property of defects is their interaction with curvature. There are actually three distinct versions of the xy model in a curved environment, which are illustrated schematically in Fig. 6.6. Here, we discuss each of these versions and emphasize the differences among them.

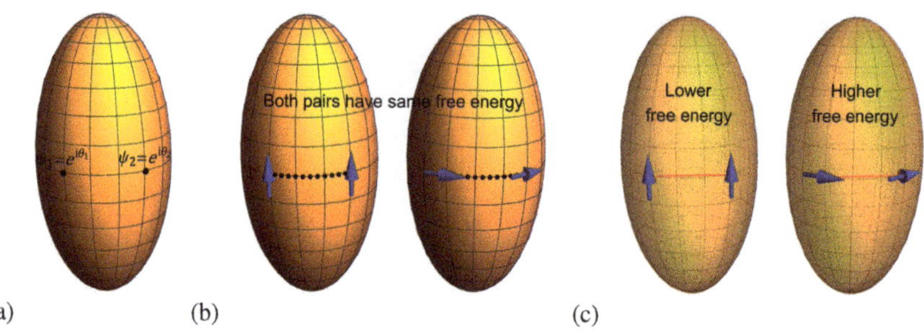

Fig. 6.6 Schematic illustrations of three models for the interaction between xy order and curvature. (**a**) Superfluid. (**b**) Liquid crystal or magnet, with an intrinsic-only interaction defined by parallel transport along the curved surface. (**c**) Liquid crystal or magnet, with an intrinsic+extrinsic interaction in 3D space

In all three of these versions, we have a unit vector $\hat{\boldsymbol{n}}(\boldsymbol{r})$ defined on a 2D curved surface; i.e. $\boldsymbol{r}$ is a position on a 2D curved surface. At each point on the surface, the curvature is characterized by a curvature tensor $\boldsymbol{K}(\boldsymbol{r})$, which has eigenvalues of $1/r_1$ and $1/r_2$, the two principal curvatures at that point. The product $G(\boldsymbol{r}) = \det(\boldsymbol{K}(\boldsymbol{r})) = 1/(r_1 r_2)$ is called the Gaussian curvature. A sphere is an example of a shape with positive Gaussian curvature, while a saddle is a shape with negative Gaussian curvature.

6.3.1 Superfluid

The *first* version of the xy model is a physical system like a superfluid. In this version, as we already discussed in Sect. 6.1.1, the xy order does not point in any direction in real space. Instead, it points some direction in an abstract order parameter space. It might be written as a unit vector $\hat{\boldsymbol{n}}(\boldsymbol{r}) = (\cos\theta(\boldsymbol{r}), \sin\theta(\boldsymbol{r}))$ or as a complex number $\psi(\boldsymbol{r}) = e^{i\theta(\boldsymbol{r})}$; these are two equivalent ways of representing the same mathematical object.

Figure 6.6a illustrates this type of order defined on a curved surface. In this figure, note that ψ is represented by a dot, not by an arrow, because it does not point in any direction with respect to the curved surface in real space.

This model has been studied in a few theoretical papers.[8] Those researchers showed that the free energy of Eq. (3.1) generalizes to

$$F = \frac{1}{2}K \int d\boldsymbol{r} |\nabla\hat{\boldsymbol{n}}|^2 = \frac{1}{2}K \int d\boldsymbol{r} |\nabla\psi|^2 = \frac{1}{2}K \int d\boldsymbol{r} |\nabla\theta|^2. \tag{6.34}$$

[8] V. Vitelli and A. M. Turner, "Anomalous Coupling Between Topological Defects and Curvature," *Phys. Rev. Lett.* **93**, 215301 (2004); A. M. Turner, V. Vitelli, and D. R. Nelson, "Vortices on Curved Surfaces," *Rev. Mod. Phys.* **82**, 1301 (2010).

In this expression, the measure of integration $d\boldsymbol{r}$ implicitly includes a factor of $\sqrt{g}$, where $g = \det(g_{ij})$ is the determinant of the local covariant metric tensor. This measure of integration provides some small coupling between xy order and curvature, but it is a fairly weak coupling compared with the following two models. This coupling affects defects, because it changes the area inside the core radius r_{core}. Because of the coupling, a defect is repelled by regions of positive Gaussian curvature, and attracted to regions of negative Gaussian curvature. The strength of the repulsion or attraction is proportional to q^2, where q is the topological charge of the defect. Hence, it affects positive- and negative-charged defects equally.

6.3.2 Liquid Crystal or Magnet with Intrinsic-Only Interaction

The *second and third* versions of the xy model both involve a liquid crystal or a magnet on a 2D curved surface. In these versions, the unit vector $\hat{\boldsymbol{n}}(\boldsymbol{r})$ points in some direction in real space, as discussed in Sect. 6.1.2. To be specific, $\hat{\boldsymbol{n}}(\boldsymbol{r})$ is a vector in the local tangent plane, i.e. the plane that is tangent to the curved surface at position $\boldsymbol{r}$. Hence, it is represented by an arrow in Fig. 6.6b and c.

For the *topology* of xy order on the surface, we must consider the Poincaré-Hopf theorem. This theorem states that the total topological charge of all defects on the entire surface must equal the Euler characteristic of the surface. For a sphere, ellipsoid, or any other closed, smooth surface with no holes, the Euler characteristic is 2. For a torus, or any other closed, smooth surface with one hole, the Euler characteristic is 0. Hence, a sphere might have two defects of $q = +1$ and no defects of $q = -1$, or three defects of $q = +1$ and one defect of $q = -1$, or any other possibility with the same total topological charge. We might say that xy order on a sphere experiences *geometric frustration*, because it cannot be defined everywhere without defects. We will discuss the concept of geometric frustration further in Chap. 12.

For the *free energy* of xy order on the surface, we need to compare the orientations $\hat{\boldsymbol{n}}(\boldsymbol{r}_1)$ and $\hat{\boldsymbol{n}}(\boldsymbol{r}_2)$ at nearby positions $\boldsymbol{r}_1$ and $\boldsymbol{r}_2$. The free energy should be low if these orientations are well aligned, and high if they are poorly aligned. The question is: *How* can we compare these orientations? That question is actually quite subtle, considering that the surface is curved, and each orientation is defined in a *different* local tangent plane.

For the *second* version of the xy model, we compare the orientations using the mathematical method of *parallel transport*. To visualize this method, imagine that a group of ants is studying $\hat{\boldsymbol{n}}(\boldsymbol{r})$. The ants can crawl on the curved surface, and they are unaware of anything off of the surface. To compare the orientations at two positions, the ants pick up the unit vector $\hat{\boldsymbol{n}}(\boldsymbol{r}_1)$ and carry it along the geodesic line from $\boldsymbol{r}_1$ to $\boldsymbol{r}_2$, maintaining a constant angle between the unit vector and the geodesic. Once they arrive at $\boldsymbol{r}_2$, they can see how well the transported version of $\hat{\boldsymbol{n}}(\boldsymbol{r}_1)$ is aligned with $\hat{\boldsymbol{n}}(\boldsymbol{r}_2)$. This comparison is illustrated schematically by the dotted path in Fig. 6.6b. In both examples in this figure, the vectors are perfectly aligned as defined by parallel transport.

There is a large theoretical literature investigating the xy model on 2D curved surfaces, with a free energy defined through parallel transport.[9] In this literature, the director field is represented as $\hat{\boldsymbol{n}}(\boldsymbol{r}) = \hat{\boldsymbol{a}}(\boldsymbol{r})\cos\theta(\boldsymbol{r}) + \hat{\boldsymbol{b}}(\boldsymbol{r})\sin\theta(\boldsymbol{r})$, where $\hat{\boldsymbol{a}}(\boldsymbol{r})$ and $\hat{\boldsymbol{b}}(\boldsymbol{r})$ form an orthonormal basis for the local tangent plane, both orthogonal to the unit normal vector $\hat{\boldsymbol{N}}(\boldsymbol{r})$. When we take derivatives of the director field, we must calculate derivatives of these basis vectors as well as derivatives of $\theta(\boldsymbol{r})$. Derivatives of the basis vectors can be described by the connection $A_i = \hat{\boldsymbol{a}} \cdot \partial_i \hat{\boldsymbol{b}} = -\hat{\boldsymbol{b}} \cdot \partial_i \hat{\boldsymbol{a}}$. Remarkably, this connection is related to the Gaussian curvature G by $\mathrm{curl}(\boldsymbol{A}) = G$. Derivatives of the director field then become

$$\nabla\hat{\boldsymbol{n}} = \underbrace{(\nabla\theta - \boldsymbol{A})(\hat{\boldsymbol{b}}\cos\theta - \hat{\boldsymbol{a}}\sin\theta)}_{(\nabla\hat{\boldsymbol{n}})_{\text{intrinsic}}} + \underbrace{(\boldsymbol{K}\cdot\hat{\boldsymbol{n}})\hat{\boldsymbol{N}}}_{(\nabla\hat{\boldsymbol{n}})_{\text{extrinsic}}}. \tag{6.35}$$

Note that it breaks up into an *intrinsic* component in the local tangent plane and an *extrinsic* component in the local normal direction, and the extrinsic component involves the curvature tensor $\boldsymbol{K}$. For parallel transport, only the intrinsic component is relevant. In that case, the free energy of Eq. (3.1) generalizes to

$$F = \frac{1}{2}K\int d\boldsymbol{r}(\nabla\hat{\boldsymbol{n}})^2_{\text{intrinsic}} = \frac{1}{2}K\int d\boldsymbol{r}|\nabla\theta - \boldsymbol{A}|^2. \tag{6.36}$$

This free energy is similar to the free energy of a superconducting phase angle in a magnetic vector potential, whose curl is the magnetic field.

Through the vector field $\boldsymbol{A}$, this model has a strong coupling between xy order and Gaussian curvature. Because of this coupling, a defect of positive topological charge is attracted to regions of positive Gaussian curvature, and repelled from regions of negative Gaussian curvature. Likewise, a defect of negative topological charge is attracted to regions of negative Gaussian curvature, and repelled from regions of positive Gaussian curvature. This coupling works in both directions: Gaussian curvature influences the positions of defects, and defects favor shapes with local Gaussian curvature.

Thus, in this model, defects interact with curvature in two ways. First, topology provides a global constraint on the total defect charge, which must match the Euler characteristic of the surface. Second, the free energy provides a local coupling between defects and Gaussian curvature, which affects both defect positions and membrane shape.

[9] The statistical mechanics of flexible membranes with orientational order was explored by D. R. Nelson and L. Peliti, "Fluctuations in Membranes with Crystalline and Hexatic Order," *Journal de Physique* **48**, 1085 (1987). The coupling between defects and Gaussian curvature was studied by J.-M. Park and T. C. Lubensky, "Topological Defects on Fluctuating Surfaces: General Properties and the Kosterlitz-Thouless Transition," *Phys. Rev. E* **53**, 2648 (1996). Many other papers have explored other aspects of this problem. Much of this literature uses the term *hexatic* order, which is a generalization of xy order with sixfold rather than onefold symmetry at each position. We will discuss hexatic order in Sect. 9.4.2.

Derivation

Here is a brief sketch of how to derive the coupling between topological defects and Gaussian curvature, for any readers who might be interested:

In the free energy of Eq. (6.36), consider the combined vector field $\boldsymbol{v}(\boldsymbol{r}) = \nabla\theta - \boldsymbol{A}$. Like any vector field, $\boldsymbol{v}(\boldsymbol{r})$ can be decomposed into longitudinal and transverse parts, $\boldsymbol{v}(\boldsymbol{r}) = \boldsymbol{v}^L(\boldsymbol{r}) + \boldsymbol{v}^T(\boldsymbol{r})$. The longitudal part has nonzero divergence and zero curl, while the transverse part has nonzero curl and zero divergence. Hence, the free energy breaks up into longitudal and transverse parts,

$$F = \frac{1}{2}K \int d\boldsymbol{r} |\boldsymbol{v}^L|^2 + \frac{1}{2}K \int d\boldsymbol{r} |\boldsymbol{v}^T|^2. \tag{6.37}$$

Let us think about each part separately.

Longitudinal The longitudinal part of $\boldsymbol{A}$ can be written as the gradient of some function $\Theta(\boldsymbol{r})$, so that $\boldsymbol{A}^L = \nabla\Theta$. The longitudinal part of $\boldsymbol{v}$ then becomes $\boldsymbol{v}^L = \nabla(\theta - \Theta)$. Hence, $\Theta(\boldsymbol{r})$ just shifts the baseline for $\theta(\boldsymbol{r})$. The minimum free energy state on this curved surface is $\theta(\boldsymbol{r}) = \Theta(\boldsymbol{r}) + \text{constant}$, instead of $\theta(\boldsymbol{r}) = \text{constant}$. This is a nonsingular change in $\theta(\boldsymbol{r})$, so it does not involve defects.

Transverse For the transverse part of $\boldsymbol{v}$, we already know the curl of each term. The curl of the first term is the topological charge density, $\text{curl}(\nabla\theta) = 2\pi\sigma$. The curl of the second term is the Gaussian curvature, $\text{curl}(\boldsymbol{A}) = G$. The total curl is $\text{curl}(\boldsymbol{v}^T) = 2\pi\sigma - G$. Hence, the topological charge density combines with the Gaussian curvature in the transverse part of the free energy. If the topological charge and the Gaussian curvature have the same sign, then they partially cancel each other, and give a reduced free energy. If the topological charge and the Gaussian curvature have opposite signs, then they add to each other, and give an increased free energy.

6.3.3 Liquid Crystal or Magnet with Intrinsic+Extrinsic Interaction

Parallel transport is not the only way to compare the orientations $\hat{\boldsymbol{n}}(\boldsymbol{r}_1)$ and $\hat{\boldsymbol{n}}(\boldsymbol{r}_2)$ at nearby positions $\boldsymbol{r}_1$ and $\boldsymbol{r}_2$. For the *third* version of the xy model, we can consider these orientations as full 3D vectors. Each orientation is confined to the local tangent plane, but the local tangent planes are embedded in the same 3D space. To visualize this comparison, imagine that a group of bees is studying $\hat{\boldsymbol{n}}(\boldsymbol{r})$. The bees are fully aware of all three dimensions, so they can just calculate the conventional 3D dot product $\hat{\boldsymbol{n}}(\boldsymbol{r}_1) \cdot \hat{\boldsymbol{n}}(\boldsymbol{r}_2)$, and hence determine how well the vectors are aligned in 3D. This comparison is illustrated schematically by the solid red line in Fig. 6.6c. Based on this criterion, the vectors in the left-hand example are better aligned in 3D, and hence have a lower free energy, than the vectors in the right-hand example.

Using this 3D comparison, the free energy of Eq. (3.1) generalizes to

$$F = \frac{1}{2}K \int dr (\nabla \hat{\boldsymbol{n}})^2 = \frac{1}{2}K \int dr \left[(\nabla \hat{\boldsymbol{n}})^2_{\text{intrinsic}} + (\nabla \hat{\boldsymbol{n}})^2_{\text{extrinsic}} \right]$$

$$= \frac{1}{2}K \int dr \big[\underbrace{|\nabla\theta - \boldsymbol{A}|^2}_{\text{intrinsic}} + \underbrace{\hat{\boldsymbol{n}} \cdot \boldsymbol{K} \cdot \boldsymbol{K} \cdot \hat{\boldsymbol{n}}}_{\text{extrinsic}} \big]. \tag{6.38}$$

This free energy has both intrinsic and extrinsic terms, and they each have the same coefficient. The intrinsic term is exactly the same as in Sect. 6.3.2, and it has the same effect: Defects of positive topological charge are attracted to regions of positive Gaussian curvature, and repelled from regions of negative Gaussian curvature, and vice versa for defects of negative topological charge. The extrinsic term has a different effect: It tends to align the director $\hat{\boldsymbol{n}}$ in the *least-curved* direction on the surface, so that the unit vectors can be aligned as well as possible in 3D. It looks similar to a magnetic field acting on the director, as in Eq. (3.17), except that it is quadratic rather than linear in $\hat{\boldsymbol{n}}$. The extrinsic term involves the full director field and the full curvature tensor. Its effects cannot be summarized in terms of topological charge and Gaussian curvature.

The model with intrinsic+extrinsic interactions has not been studied in very many theoretical papers,[10] perhaps because the extrinsic term is less universal than the intrinsic term.[11] However, there is no reason to believe that the extrinsic term is small. If it is present in any experimental system, then it should be just as important as the intrinsic term.

Example

As a specific example to illustrate the difference between the intrinsic-only model and the intrinsic+extrinsic model, consider an xy model on a cylinder. Note that $\hat{\boldsymbol{n}}(\boldsymbol{r})$ is defined on the 2D cylindrical surface, *not* in the 3D interior of the cylinder. Suppose the cylinder has fixed radius ρ, and it is parameterized by the azimuthal angle ϕ (from $-\pi$ to π) and height z (from $-h$ to h). The director is $\hat{\boldsymbol{n}} = \hat{\boldsymbol{\phi}} \cos\theta + \hat{\boldsymbol{z}} \sin\theta$. Suppose it has boundary conditions such that $\theta = -\pi/2$ at $z = -h$, and $\theta = \pi/2$ at $z = h$.

(continued)

[10] For one study of this model, with visualizations of the resulting defect configurations, see R. L. B. Selinger, A. Konya, A. Travesset, and J. V. Selinger, "Monte Carlo Studies of the XY Model on Two-Dimensional Curved Surfaces," *J. Phys. Chem. B* **115**, 13989 (2011).

[11] The extrinsic term can only occur for xy order (with onefold symmetry at each position) or nematic order (twofold symmetry). It cannot occur for hexatic order (sixfold symmetry), or for any symmetry greater than twofold. These higher-symmetry models have only the intrinsic free energy.

The intrinsic-only model is fairly simple in this geometry. The connection is $A_i = 0$, and Gaussian curvature is $G = 0$. Hence, the intrinsic-only free energy is

$$F_{\text{int}} = \frac{1}{2}K \int \rho d\phi dz \left[\frac{1}{\rho^2} \left(\frac{\partial \theta}{\partial \phi} \right)^2 + \left(\frac{\partial \theta}{\partial z} \right)^2 \right]. \tag{6.39}$$

The Euler-Lagrange equation corresponding to this free energy is just $\rho^{-2}(\partial^2\theta/\partial\phi^2) + (\partial^2\theta/\partial z^2) = 0$, and the solution is $\theta = (\pi z)/(2h)$. This solution is illustrated in Fig. 6.7a. We can see that the variation of $\hat{\boldsymbol{n}}$ is spread out as much as possible along the length of the cylinder.

For the intrinsic+extrinsic model, the free energy is

$$F_{\text{int+ext}} = \frac{1}{2}K \int \rho d\phi dz \left[\frac{1}{\rho^2} \left(\frac{\partial \theta}{\partial \phi} \right)^2 + \left(\frac{\partial \theta}{\partial z} \right)^2 + \frac{\cos^2\theta}{\rho^2} \right]. \tag{6.40}$$

The final, extrinsic term favors $\theta = \pm\pi/2$, which keeps the director aligned with the long axis of the cylinder. The Euler-Lagrange equation becomes

$$\frac{1}{\rho^2}\frac{\partial^2\theta}{\partial\phi^2} + \frac{\partial^2\theta}{\partial z^2} + \frac{\sin 2\theta}{2\rho^2} = 0. \tag{6.41}$$

This differential equation is very similar to the sine-Gordon equation (3.19) for solitons. The solution is

$$\theta = 2\tan^{-1}\left[\tanh\left(\frac{z}{\rho} \right) \right], \tag{6.42}$$

assuming that $h \gg \rho$. This solution is illustrated in Fig. 6.7b. In this case, the variation of $\hat{\boldsymbol{n}}$ is concentrated in a belt of width ρ. Hence, we might say that the extrinsic free energy leads to a soliton in $\hat{\boldsymbol{n}}$ on the cylinder, just as the magnetic field leads to a soliton in $\hat{\boldsymbol{n}}$ in Sect. 3.3.

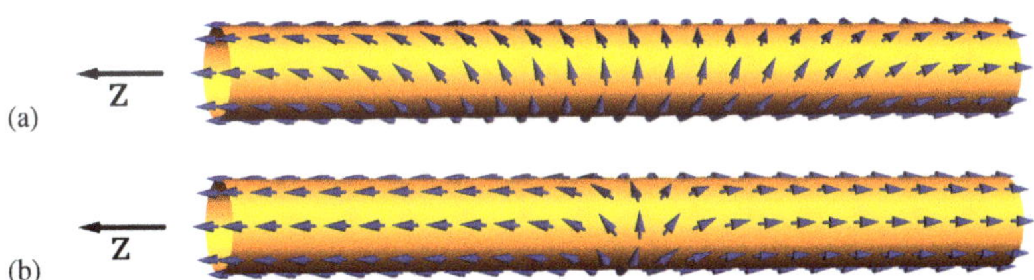

Fig. 6.7 Two models for xy order on a cylinder, with boundary conditions on the two ends. (**a**) In the intrinsic-only model, the variation of $\hat{\boldsymbol{n}}(\boldsymbol{r})$ is spread out as much as possible. (**b**) In the intrinsic+extrinsic model, the variation is concentrated in a soliton, with a width comparable to the cylinder radius

Part II

Generalizations

2D Nematic Order, Active Liquid Crystals 7

In Part I, we discussed a single type of order parameter: the xy order parameter, also known as 2D polar order. We analyzed topological defects and solitons of this order parameter using a single type of measuring surface: a 1D path. In Part II, we will generalize the discussion to other types of order parameters in Chaps. 7–9, and to other types of measuring surfaces in Chaps. 10 and 11.

In this chapter, we begin the generalization by considering 2D nematic order. Defects and solitons in 2D nematic order are fairly similar to defects and solitons in 2D polar order, but we will point out some differences. We also describe more complex defect structures that can form if a system has a combination of nematic and polar order. Finally, we consider two further issues—unequal elastic constants and active materials. These two issues are not necessarily associated with nematic order, but they often occur together with nematic order, and hence it is appropriate to discuss them in this chapter.

7.1 Half-Integer Defects and Solitons

Let us begin by distinguishing between different types of orientational order. Suppose we have an ensemble of microscopic objects, and each object has an orientation in the 2D plane, as indicated by the brown arrows on the left side of Fig. 7.1. These microscopic objects might be magnetic spins, or liquid crystal molecules, or something else. We are interested in the statistical distribution of orientations. This distribution might be isotropic, completely disordered, so that the brown arrows are equally likely to point in any direction in the 2D plane. Alternatively, the distribution might be anisotropic, so that the distribution of orientations is biased in some way.

Figure 7.1a shows one type of orientational order. On the left side, in a microscopic view, the brown arrows are most likely to point upward in the $+\hat{y}$ direction. For that reason, we can represent the average order in the local region

J. V. Selinger, *Introduction to Topological Defects and Solitons*, Lecture Notes in Physics 1032, https://doi.org/10.1007/978-3-031-70200-6_7

by a large blue arrow. This situation is exactly the xy model that we have studied in Part I. It can also be called *dipolar* order, or just *polar* order for short, because the statistical average orientation is some kind of dipole moment, perhaps a magnetic dipole. On the right side, in a more macroscopic view, each blue arrow represents the coarse-grained average orientational order in its own local region. The local average varies gradually as a function of position across the system. That variation is represented by the field $\hat{\boldsymbol{n}}(\boldsymbol{r})$.

Figure 7.1b shows a different type of orientational order. On the left side, in a microscopic view, the brown arrows are more likely to point along the $\pm\hat{y}$ axis, but they are equally likely to point up or down along this axis. Hence, we can represent the average order in the local region by a large blue *double-headed* arrow. This type of orientational order is called *nematic* order. It can also be called *quadrupolar*

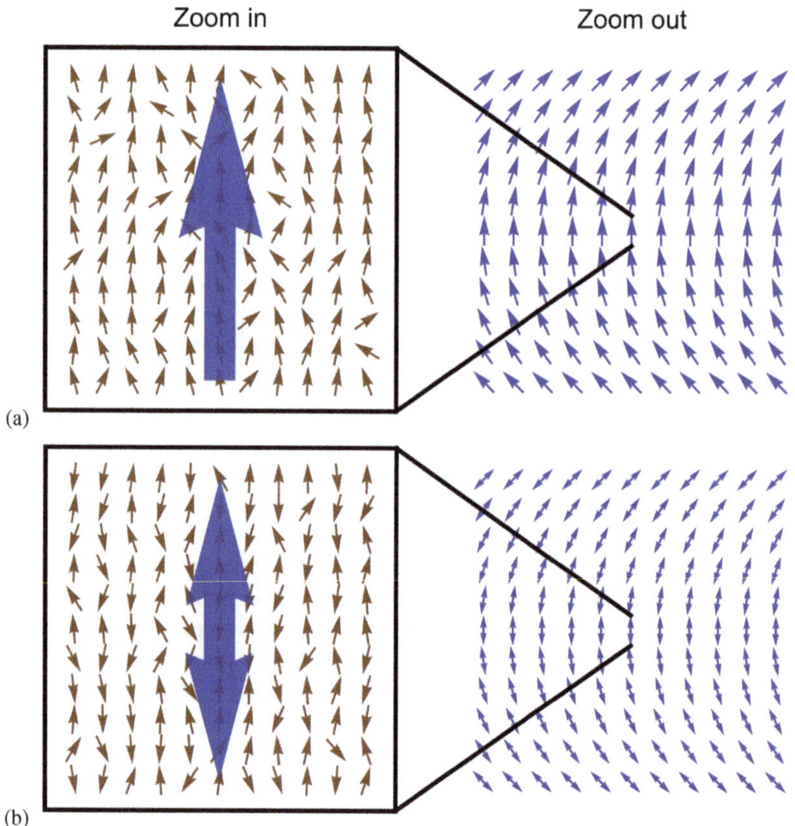

Fig. 7.1 Two different types of 2D orientational order: (**a**) Polar or xy order. (**b**) Nematic order. On the left, the zoomed-in views show the distribution of microscopic orientations. On the right, the zoomed-out views show how the coarse-grained average orientation varies as a function of position across the system. Polar or xy order is represented by a single-headed arrow, because the orientational distribution is biased in one direction. Nematic order is represented by a double-headed arrow, because the orientational distribution is biased up or down along one axis

order, because the statistical average orientation is a quadrupole moment. On the right side, in the more macroscopic view, each blue double-headed arrow represents the average nematic order in the local region. As in the previous case, the local average varies gradually as a function of position. That variation can be represented by a unit vector field $\hat{\boldsymbol{n}}(\boldsymbol{r})$, called the nematic director, as long as we understand implicitly that $\hat{\boldsymbol{n}}$ and $-\hat{\boldsymbol{n}}$ represent the same physical state.

Polar order commonly occurs in magnetism, while nematic order commonly occurs in liquid crystals. However, there are exceptions to this general rule; certain types of liquid crystals can have polar order.

With this background, we can now consider topological defects in a 2D nematic phase. As discussed in Chap. 1, a topological defect is a singularity with a winding number: If we travel around a closed measuring loop that encloses the defect, the orientational order rotates and eventually comes back to its original state. In a *polar* phase, the vector field $\hat{\boldsymbol{n}}(\boldsymbol{r})$ must rotate through a full circle to come back to the original state. However, in a *nematic* phase, $\hat{\boldsymbol{n}}(\boldsymbol{r})$ does not need to rotate through a full circle. Rather, it might only rotate through a half-circle, so that $\hat{\boldsymbol{n}}$ transforms to $-\hat{\boldsymbol{n}}$, because $\hat{\boldsymbol{n}}$ and $-\hat{\boldsymbol{n}}$ represent the same physical state of the material.

Figure 7.2 shows several examples of defect structures that can form in a 2D nematic phase. In Fig. 7.2a, as we travel along the green loop counter-clockwise around the defect, $\hat{\boldsymbol{n}}(\boldsymbol{r})$ rotates through a half-circle counter-clockwise, so that the angle $\theta(\boldsymbol{r})$ increases by π. Hence, the topological charge of this defect is $q = \frac{1}{2\pi}\oint d\theta = +1/2$. By comparison, in Fig. 7.2b, as we travel in a full loop counter-clockwise around the defect, $\hat{\boldsymbol{n}}(\boldsymbol{r})$ rotates through a half-circle clockwise, so that the angle $\theta(\boldsymbol{r})$ decreases by π. Hence, the topological charge is $q = \frac{1}{2\pi}\oint d\theta = -1/2$.

To be sure, the director field is not required to rotate through a half-circle around a defect; it could rotate through a full circle. Figure 7.2c and d shows two examples in which $\hat{\boldsymbol{n}}(\boldsymbol{r})$ rotates through a full circle counter-clockwise as we travel around the loop counter-clockwise, and hence $\theta(\boldsymbol{r})$ increases by 2π. These examples both have $q = +1$. Similarly, Fig. 7.2e is an example where $\hat{\boldsymbol{n}}(\boldsymbol{r})$ rotates through a full circle clockwise, so that $\theta(\boldsymbol{r})$ decreases by 2π. It has topological charge $q = -1$.

Higher topological charges are also possible. For example, in Fig. 7.2f, as we travel in a full loop counter-clockwise around the defect, $\hat{\boldsymbol{n}}(\boldsymbol{r})$ rotates through one and a half circles counter-clockwise, and hence $\theta(\boldsymbol{r})$ increases by 3π. This case has $q = 3/2$. In general, for a 2D nematic phase, the topological charge q can be any *half-integer or integer*, so that the orientational order will return to the same physical state after we travel around a loop.

A nematic phase can have more than one topological defect. Figure 7.3a shows an example with two defects, which each have $q = +1/2$, as can be seen by going around the small green loops. The total topological charge is $q = +1$, going around the large green loop. Likewise, Fig. 7.3b shows an example with two defects of $q = +1/2$ and $-1/2$, respectively. The total topological charge is $q = 0$.

The defect points for 2D nematic order in a 2D system can be extended into defect lines for 2D nematic order in a 3D system, analogous to the defect lines presented in Sect. 1.3. For example, Fig. 7.4a shows a straight defect line with topological charge $q = +1/2$. Defect lines can curve, and they can form closed

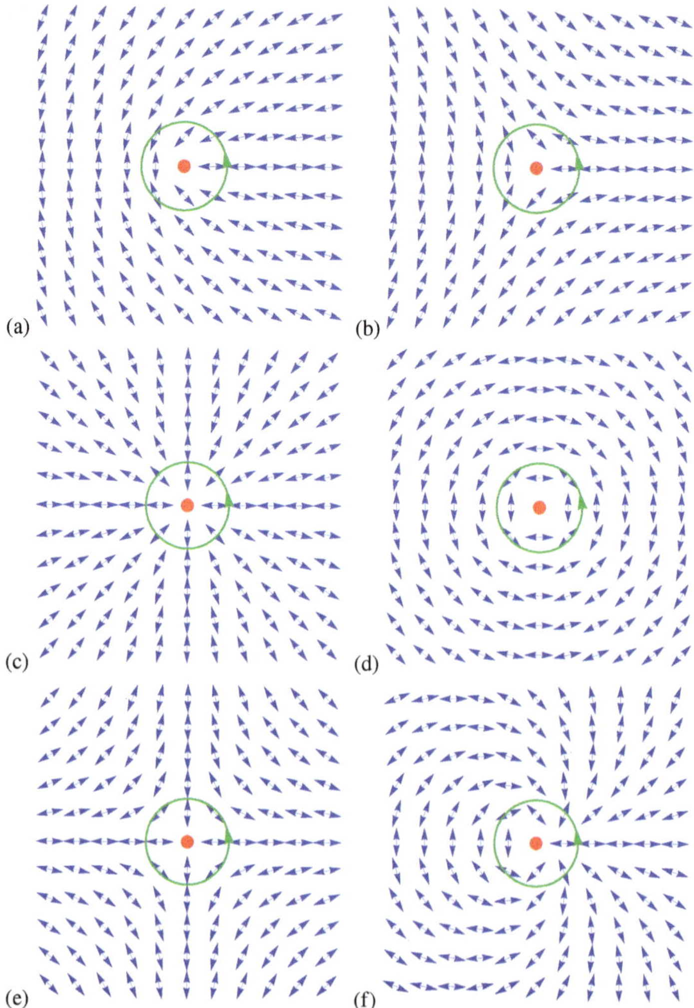

Fig. 7.2 Examples of topological defects in a 2D nematic phase. (**a**) Topological charge $q = +1/2$. (**b**) Topological charge $q = -1/2$. (**c,d**) Topological charge $q = +1$. (**e**) Topological charge $q = -1$. (**f**) Topological charge $q = +3/2$

loops, as Fig. 7.4b. Here, the small green loops on the right enclose a charge of $q = +1/2$, while the small green loops on the left enclose $q = -1/2$. The large green loops enclose a total of $q = 0$. These figures can be compared with Fig. 1.6a and d for 2D polar order.

The argument about half-integer winding number applies to topological solitons, as well as topological defects, in a phase with 2D nematic order. As discussed in Chap. 2, a topological soliton is a narrow region with a winding number: If we travel across that region, the orientational order rotates and eventually comes back to its

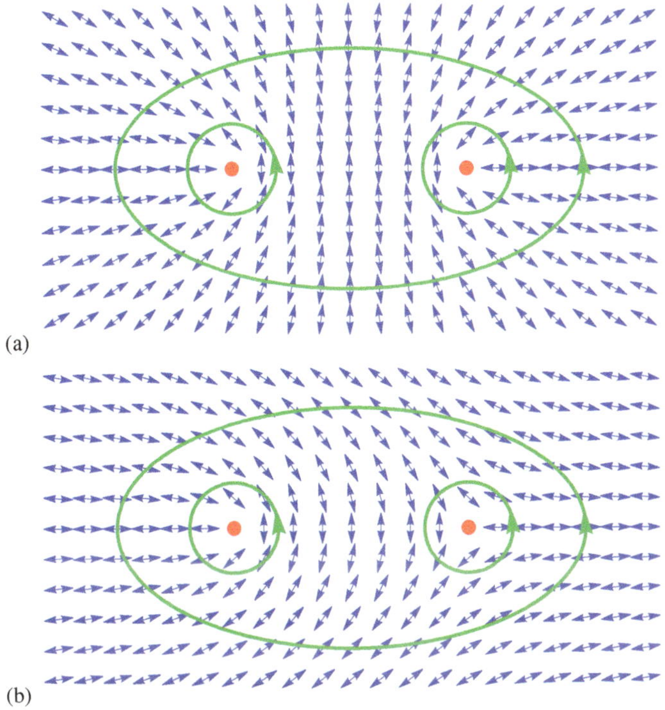

Fig. 7.3 Examples of 2D nematic director configurations with two topological defects. (**a**) Two topological charges of $+1/2$ each, which combine to make a total of $+1$. (**b**) Two topological charges of $+1/2$ and $-1/2$, which combine to make a total of 0

original state. In a *polar* phase, $\hat{\boldsymbol{n}}(x)$ must rotate through a full circle to come back to the original state. However, in a *nematic* phase, $\hat{\boldsymbol{n}}(x)$ might only rotate through a half-circle, so that $\hat{\boldsymbol{n}}$ transforms to $-\hat{\boldsymbol{n}}$, because $\hat{\boldsymbol{n}}$ and $-\hat{\boldsymbol{n}}$ represent the same physical state.

Figure 7.5a shows an example of a topological soliton in a nematic phase. As we cross a narrow width of order ξ, $\hat{\boldsymbol{n}}(x)$ rotates through a half-circle counter-clockwise, so that $\theta(x)$ increases by π. This soliton has a winding number of $q = +1/2$. By comparison, in Fig. 7.5b, $\hat{\boldsymbol{n}}(x)$ rotates through a half-circle clockwise, so that $\theta(x)$ decreases by π, and the winding number is $q = -1/2$. Both of these examples can be called π-walls, as opposed to the 2π-walls in Chap. 2. The $+1/2$ and $-1/2$ solitons can be regarded as a wall and an antiwall, because they can annihilate each other.

In most respects, topological defects and solitons in 2D nematic order are similar to topological defects and solitons in 2D polar order, except that q can be a half-integer or integer in the nematic case, while it can only be an integer in the polar case. Here, we will just mention a few points of comparison.

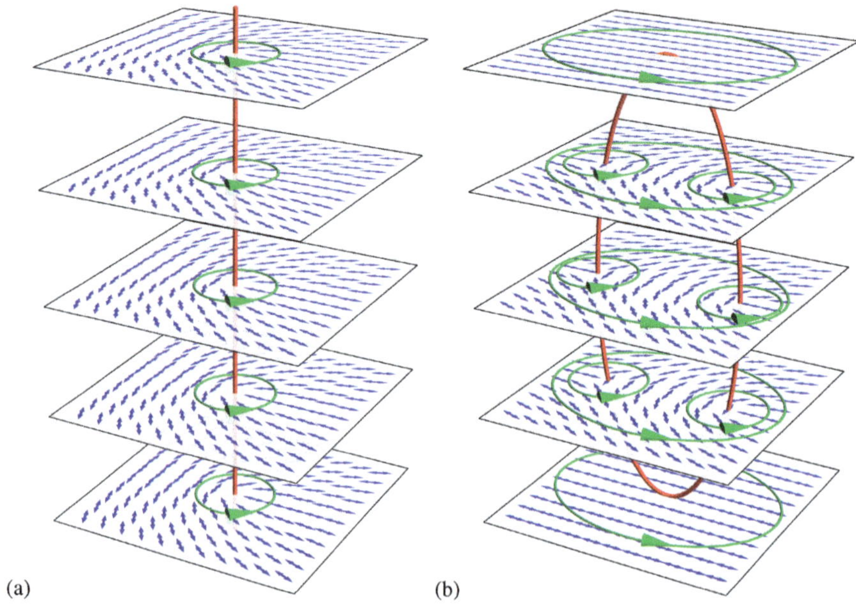

Fig. 7.4 Configurations of the 2D nematic director $\hat{\boldsymbol{n}}(\boldsymbol{r})$ in a 3D system. (**a**) Straight defect line with topological charge $+1/2$. (**b**) Defect loop

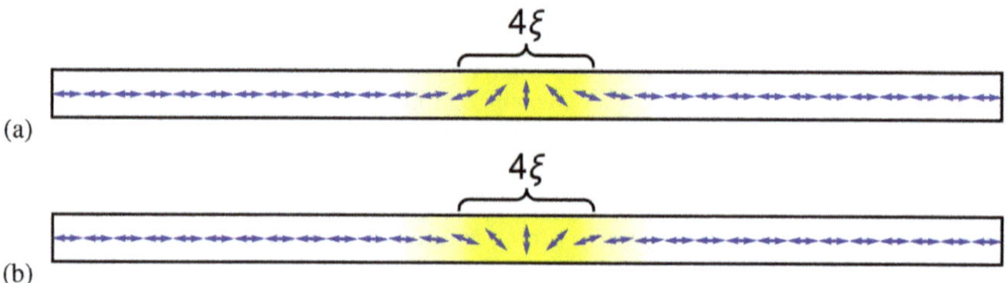

Fig. 7.5 Examples of solitons, or π-walls, in a nematic phase. (**a**) Winding number $q = +1/2$. (**b**) Winding number $q = -1/2$

7.1.1 Free Energy of Defect

In a nematic phase, as in a polar phase, spatial variations of the director $\hat{\boldsymbol{n}}(\boldsymbol{r})$ cost free energy. The simplest model assumes that the free energy density is proportional to the square of the spatial gradient $\nabla\hat{\boldsymbol{n}}$, as in Eq. (3.1), so that

$$F = \frac{1}{2}K\int d\boldsymbol{r}|\nabla\hat{\boldsymbol{n}}|^2 = \frac{1}{2}K\int d\boldsymbol{r}\left[(\nabla\cdot\hat{\boldsymbol{n}})^2 + |\nabla\times\hat{\boldsymbol{n}}|^2\right]$$
$$= \frac{1}{2}K\int d\boldsymbol{r}(\partial_i n_j)(\partial_i n_j) = \frac{1}{2}K\int d\boldsymbol{r}|\nabla\theta|^2. \tag{7.1}$$

In Sect. 7.3 below, we will compare this simple model with a more general model for the free energy density. As long as we keep this simple model, the discussion in Sects. 3.1 and 3.2 still applies. The free energy of a single defect is still given by $F = \pi K q^2 \log(R/r_{\text{core}})$, where R is the system size and r_{core} is the core radius, but now q can be a half-integer or integer. Because the free energy is proportional to q^2, any defect with high q will tend to break up into defects with the minimum possible q, which is $\pm 1/2$, in order to reduce the free energy. Pairs of defects still have a Coulomb-like interaction, with an interaction energy that scales logarithmically with distance r_{12}, and hence a force that scales as $1/r_{12}$.

7.1.2 Free Energy of Soliton

In Sect. 3.3, we saw that a system needs some type of aligning field, such as a magnetic field, in order to make a soliton with a characteristic width ξ. In a polar phase, the aligning energy has the form of $\Delta F = -\boldsymbol{\mu} \cdot \boldsymbol{B} = -\mu \hat{\boldsymbol{n}} \cdot \boldsymbol{B}$. In a nematic phase, the aligning energy cannot take that form. Because $\hat{\boldsymbol{n}}$ and $-\hat{\boldsymbol{n}}$ represent the same physical state, they must have the same free energy, so all terms in the free energy must be *even* functions of $\hat{\boldsymbol{n}}$. However, a nematic phase can have a diamagnetic anisotropy $\Delta\chi$, which gives an aligning energy of the form $\Delta F = -(\Delta\chi/(2\mu_0))(\hat{\boldsymbol{n}} \cdot \boldsymbol{B})^2$. Combining the elastic and magnetic contributions, the free energy becomes

$$
\begin{aligned}
F &= \int_{-\infty}^{\infty} dx \left[\frac{1}{2} K |\nabla \hat{\boldsymbol{n}}|^2 - \frac{\Delta\chi (\hat{\boldsymbol{n}} \cdot \boldsymbol{B})^2}{2\mu_0} \right] \\
&= \int_{-\infty}^{\infty} dx \left[\frac{1}{2} K \left(\frac{d\theta}{dx} \right)^2 - \frac{\Delta\chi B^2}{2\mu_0} \cos^2\theta \right].
\end{aligned} \tag{7.2}
$$

The corresponding Euler-Lagrange equation is

$$
\frac{d^2(2\theta)}{dx^2} = \frac{1}{\xi^2} \sin(2\theta), \quad \text{with } \xi = \frac{1}{B} \sqrt{\frac{\mu_0 K}{\Delta\chi}}. \tag{7.3}
$$

The length scale ξ is the magnetic coherence length of liquid crystals. This differential equation is just another form of the sine-Gordon equation, and the solution is

$$
\theta(x) = \frac{\pi}{2} \pm 2 \tan^{-1} \left[\tanh \left(\frac{x - x_0}{2\xi} \right) \right], \tag{7.4}
$$

with the $\pm$ sign for a soliton of winding number $q = \pm 1/2$. Using that solution for $\theta(x)$, the rest of the discussion in Sects. 3.3 and 3.4 still applies.

7.1.3 Variable Magnitude of Order

In Sect. 5.1, we discussed how to describe defects in a polar phase using a variable-magnitude theory. For a nematic phase, the same general concept still works, but the mathematical formalism is different. Polar order of variable magnitude is described by a vector order parameter $\boldsymbol{M}(\boldsymbol{r}) = M\hat{\boldsymbol{n}}$, but nematic order of variable magnitude is described by a tensor order parameter $Q_{ij}(\boldsymbol{r})$. In 2D, this tensor order parameter is related to the director $\hat{\boldsymbol{n}}$ and the angle θ by[1]

$$Q_{ij} = S(2n_i n_j - \delta_{ij}) = S\begin{pmatrix} 2\cos^2\theta - 1 & 2\cos\theta\sin\theta \\ 2\cos\theta\sin\theta & 2\sin^2\theta - 1 \end{pmatrix} = S\begin{pmatrix} \cos 2\theta & \sin 2\theta \\ \sin 2\theta & -\cos 2\theta \end{pmatrix}. \tag{7.5}$$

Here, $S(\boldsymbol{r})$ is a scalar that represents the magnitude of nematic order, just as $M(\boldsymbol{r})$ is a scalar representing the magnitude of polar order. The tensor Q_{ij} is symmetric and traceless; its eigenvalues are $\pm S$; and its eigenvectors are $\hat{\boldsymbol{n}}$ and $\hat{\boldsymbol{z}} \times \hat{\boldsymbol{n}}$. Using the same style of theory as in Chap. 5, we can write the free energy as

$$F = \int d\boldsymbol{r}\left[-\frac{1}{4}aQ_{ij}Q_{ij} + \frac{1}{16}b(Q_{ij}Q_{ij})^2 + \frac{1}{4}L(\partial_k Q_{ij})(\partial_k Q_{ij})\right] \tag{7.6}$$

$$= \int d\boldsymbol{r}\left[-\frac{1}{2}aS^2 + \frac{1}{4}bS^4 + \frac{1}{2}L|\nabla S|^2 + 2LS^2|\nabla\theta|^2\right], \tag{7.7}$$

summing over repeated indices with the Einstein summation convention. The corresponding Euler-Lagrange equations for $S(\boldsymbol{r})$ and $\theta(\boldsymbol{r})$ are

$$\frac{\delta F}{\delta S(\boldsymbol{r})} = -aS + bS^3 - L\nabla^2 S + 4LS|\nabla\theta|^2 = 0, \tag{7.8a}$$

$$\frac{\delta F}{\delta\theta(\boldsymbol{r})} = -4LS^2\nabla^2\theta - 8LS(\nabla S)\cdot(\nabla\theta) = 0. \tag{7.8b}$$

To describe a defect of topological charge q, we can use our solution $\theta = q\phi + \theta_0$ in polar coordinates (r, ϕ). Hence, S must be a function of r alone, independent of ϕ, satisfying

$$-aS + bS^3 - L\frac{d^2S}{dr^2} - \frac{L}{r}\frac{dS}{dr} + \frac{4Lq^2}{r^2}S = 0. \tag{7.9}$$

[1] This expression for Q_{ij} describes nematic order in a 2D plane. It differs from the standard expression for nematic order in 3D space, which will be discussed in Sect. 8.4.4. For further background on the tensor order parameter for a nematic phase, see our previous book J. V. Selinger, *Introduction to the Theory of Soft Matter: From Ideal Gases to Liquid Crystals* (Springer, 2016), Section 10.2.

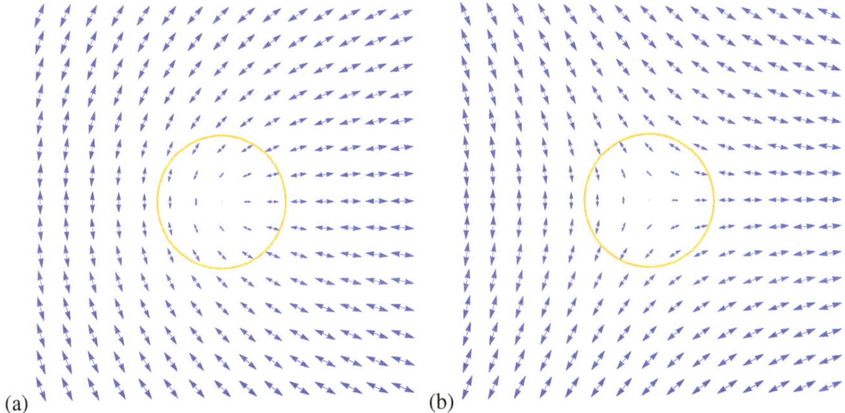

Fig. 7.6 Visualization of topological defects in a 2D nematic phase, using the variable-magnitude theory. The direction of the double-headed arrows represents the director field $\hat{\boldsymbol{n}}(\boldsymbol{r})$, and the length of these arrows represents the magnitude $S(r)$. The gold circles show the radius ρ_{core} around each defect. (**a**) $q = +1/2$. (**b**) $q = -1/2$

This equation (with $q = \pm 1/2$) is exactly the same as Eq. (5.10) (with $q = \pm 1$). Hence, it has the same solution, which is approximately $S(r) \approx S_0(1 - e^{-0.8r/\rho_{\text{core}}})$, with the bulk magnitude $S_0 = \sqrt{a/b}$ and the core radius $\rho_{\text{core}} = \sqrt{L/a}$.

Figure 7.6a and b shows the nematic order around defects of topological charge $q = +1/2$ and $q = -1/2$, respectively. In these visualizations, the direction of the double-headed arrows represents $\hat{\boldsymbol{n}}(\boldsymbol{r})$, which gives the orientation of nematic order, and the length of these arrows represents $S(\boldsymbol{r})$, which gives the magnitude of nematic order. The gold circles show the radius ρ_{core}. These figures are analogous to Fig. 5.3a and b for polar order. Far from the defect centers, the magnitude of nematic order is approximately constant. However, close to the centers, Fig. 7.6a and b gives a more microscopic view of the defect cores, where the nematic order goes smoothly to zero.

7.1.4 Defect Phase or Orientation

In Sect. 6.1, we showed that defects in a polar phase can be regarded as objects with a phase or orientation. We characterized the orientation based on where $\hat{\boldsymbol{n}}(\boldsymbol{r})$ points radially outward. In a nematic phase, we can still use the same general concept, but with one difference: Because $\hat{\boldsymbol{n}}$ and $-\hat{\boldsymbol{n}}$ represent the same physical state, we cannot say where $\hat{\boldsymbol{n}}$ points outward; we can only say where it points radially *outward or inward*. That alignment occurs when $\theta = \phi \pmod{\pi}$, not $\pmod{2\pi}$. For a defect of topological charge q, with $\theta = q\phi + \theta_0$, the radial alignment occurs at $\theta = \phi = \psi$, where

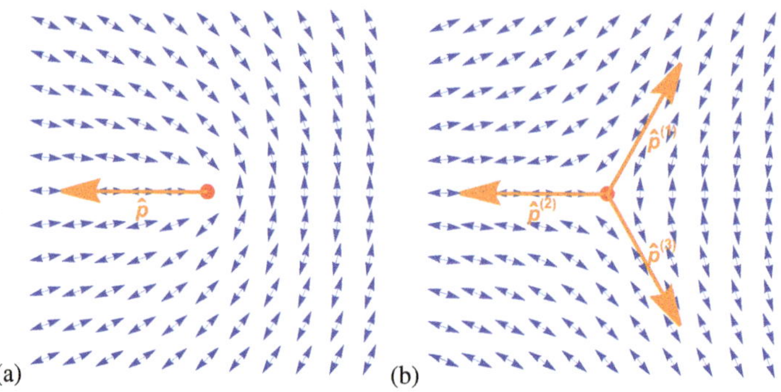

Fig. 7.7 Orientation of defects in a 2D nematic phase. (**a**) Topological charge $q = +1/2$, with onefold symmetry. (**b**) Topological charge $q = -1/2$, with threefold symmetry. Both examples have $\theta_0 = \pi/2$

$$\psi = \frac{\theta_0}{1-q} \quad \left(\text{mod}\ \frac{\pi}{|1-q|}\right). \tag{7.10}$$

This orientation can be represented as a unit vector $\hat{\boldsymbol{p}} = (\cos\psi, \sin\psi)$.

For a defect of topological charge $q = +1/2$, Eq. (6.6) becomes $\psi = 2\theta_0$ (mod 2π). Hence, the radial alignment occurs at just one angle ψ between 0 and 2π, and the defect orientation is characterized by just one unit vector $\hat{\boldsymbol{p}}$, represented by the orange arrow in Fig. 7.7a. This unit vector can be determined from the director field by[2]

$$\hat{\boldsymbol{p}} = \frac{\nabla \cdot (\hat{\boldsymbol{n}}\hat{\boldsymbol{n}})}{|\nabla \cdot (\hat{\boldsymbol{n}}\hat{\boldsymbol{n}})|}. \tag{7.11}$$

The defect structure has only onefold rotational symmetry. It is often described as *comet-shaped*, with $\hat{\boldsymbol{p}}$ pointing from the head to the tail of the comet.

By comparison, for a defect of topological charge $q = -1/2$, Eq. (6.6) becomes $\psi = 2\theta_0/3$ (mod $2\pi/3$). In this case, the radial alignment occurs at three angles ψ between 0 and 2π, and the defect orientation is characterized by three unit vectors $\hat{\boldsymbol{p}}$, indicated by the triad of three orange arrows in Fig. 7.7b. Those three vectors can be combined into a single tensor of rank three,

$$\boldsymbol{T} = \frac{2}{3}\left(\hat{\boldsymbol{p}}^{(1)}\hat{\boldsymbol{p}}^{(1)}\hat{\boldsymbol{p}}^{(1)} + \hat{\boldsymbol{p}}^{(2)}\hat{\boldsymbol{p}}^{(2)}\hat{\boldsymbol{p}}^{(2)} + \hat{\boldsymbol{p}}^{(3)}\hat{\boldsymbol{p}}^{(3)}\hat{\boldsymbol{p}}^{(3)}\right), \tag{7.12}$$

[2] See A. J. Vromans and L. Giomi, "Orientational Properties of Nematic Disclinations," *Soft Matter* **12**, 6490 (2016), and further discussion in X. Tang and J. V. Selinger, "Orientation of Topological Defects in 2D Nematic Liquid Crystals," *Soft Matter* **13**, 5481 (2017).

using the tensor or outer product. The defect structure has threefold rotational symmetry.

In general, for an arbitrary topological charge $q \neq +1$, there are $2|1-q|$ angles ψ between 0 and 2π, corresponding to $2|1-q|$ distinct unit vectors $\hat{\boldsymbol{p}}$, and the defect has $2|1-q|$-fold rotational symmetry. It might be represented by a tensor of rank $2|1-q|$, by analogy with Eq. (6.8). The topological charge of $q = +1$ is still a special case, where the phase θ_0 determines the structure rather than the orientation of a defect.

7.2 Interaction Between Nematic and Polar Order

Suppose that a phase has *strong* nematic order and *weak* polar order. For example, in Fig. 7.1, the phase might have 51% of brown arrows pointing along $\hat{\boldsymbol{n}}$, and 49% of brown arrows pointing along $-\hat{\boldsymbol{n}}$. In this case, the system forms complex string defect structures, as illustrated in Fig. 7.8.[3] There are actually two ways to think about these structures, as we will outline in this section. These two ways do not contradict each other—rather, they are two complementary perspectives about the same physical objects.

From one perspective, a phase with strong nematic order and weak polar order can just be regarded as *polar*. Polar is more ordered and less symmetric than nematic. Even a small magnitude of polar order still breaks the nematic symmetry between $\hat{\boldsymbol{n}}$ and $-\hat{\boldsymbol{n}}$. We have already analyzed defects in a polar phase, and we know that a polar phase can only have defects with *integer* values of the topological charge. Each defect in Fig. 7.8 has $q = +1$, as can be seen by drawing a large loop around the entire red object. In this way of thinking, the string structure is just a complex defect core. A defect core is usually disk-shaped, but in this particular case, it is extended in one dimension.

From another perspective, a phase with strong nematic order and weak polar order can be regarded as *almost nematic*. The nematic symmetry between $\hat{\boldsymbol{n}}$ and $-\hat{\boldsymbol{n}}$ is only slightly broken. For that reason, defect structures in this phase should be similar to defect structures in an ordinary nematic phase. An ordinary nematic phase can have a configuration with two $+1/2$ defects, as shown in Fig. 7.3a. The string defect structure in Fig. 7.8a is a slightly modified version of that configuration. The modification is that the two $+1/2$ nematic defects are connected by a domain wall in the polar order. Across the domain wall, the polar order reverses direction. We can make different microscopic models for what might happen inside the domain wall—the polar order might rapidly rotate through an angle of π (similar to the solitons that we have discussed), or its magnitude might pass through zero. If the polar order is very weak, the domain wall is wide, so that it appears fuzzy, and it

[3] D. H. Lee and G. Grinstein, "Strings in Two-Dimensional Classical XY Models," *Phys. Rev. Lett.* **55**, 541 (1985).

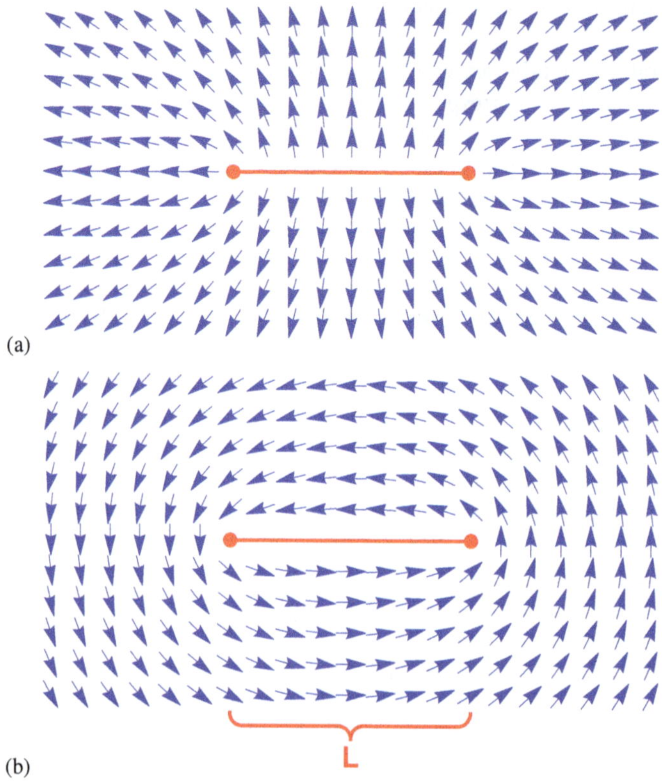

Fig. 7.8 Examples of string defects in a phase with strong nematic order and weak polar order. Each string defect consists of two $+1/2$ nematic defects connected by a domain wall in the polar order. Examples (**a**) and (**b**) are identical except for an overall rotation of the orientational order (change of the phase θ_0) by $\pi/2$

has a low energy per length ϵ_W. If the polar order grows stronger, the domain wall becomes narrower, so that it appears sharper, and it has a higher energy per length.

Based on this second perspective, we can determine the equilibrium length L of the string defect. The free energy of the string defect has two parts. First, there is the repulsion between the two $+1/2$ nematic defects, which is $2\pi K q_1 q_2 \log(L_{\max}/L)$, with $q_1 = q_2 = 1/2$. Second, there is the free energy of the domain wall itself, which is $\epsilon_W L$. Hence, the total free energy of the string defect becomes

$$F_{\text{string}} = \frac{\pi K}{2} \log\left(\frac{L_{\max}}{L}\right) + \epsilon_W L. \tag{7.13}$$

Minimizing the free energy over L gives

$$L = \frac{\pi K}{2\epsilon_W}. \tag{7.14}$$

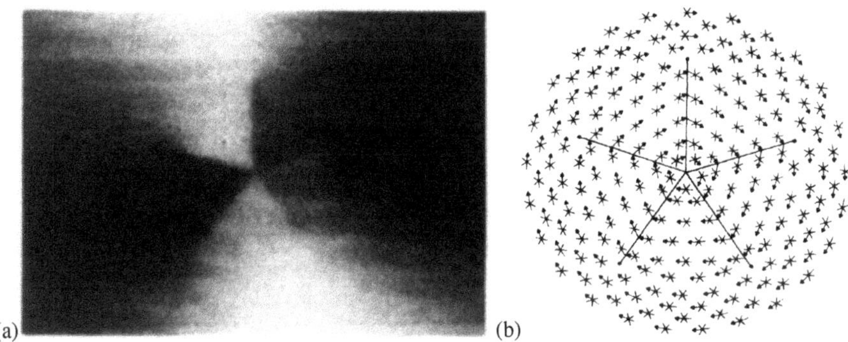

Fig. 7.9 Star defect in a thin film of a tilted hexatic liquid crystal. (**a**) Experimental image. (**b**) Model in terms of tilt order (indicated by arrows) and hexatic bond-orientational order (indicated by sixfold symmetric symbols). Reprinted with permission from S. B. Dierker, R. Pindak, and R. B. Meyer, *Phys. Rev. Lett.* **56**, 1819 (1986). Copyright 1986 by the American Physical Society

Hence, the defect length L depends on the domain wall energy ϵ_W associated with polar order. If there is no polar order, the two nematic defects are unbound, and they can move apart to $L \to \infty$. If there is weak polar order, the nematic defects are bound by a weak linear potential, and the equilibrium distance L is large but finite. As the polar order increases, the linear potential between the nematic defects grows stronger and the equilibrium L becomes smaller. Eventually, for strong polar order, the linear potential is very strong and the equilibrium L is very short. In this limit, we no longer have a string defect. Rather, the two $+1/2$ defects are pulled together into a single $+1$ defect, and the string structure crosses over to an ordinary polar defect with a disk-shaped core.

Incidentally, an even more complex version of this structure occurs in tilted hexatic liquid crystals.[4] Instead of a combination of polar order (onefold symmetry) and nematic order (twofold symmetry), a tilted hexatic phase has a combination of tilt order (onefold symmetry) and hexatic bond-orientational order (sixfold symmetry). We will discuss hexatic order in Sect. 9.4.2. This combination of order parameters gives the *star defect* shown in Fig. 7.9. From one perspective, the entire star defect can be regarded as a +1 defect in the tilt order, with a star-shaped defect core. From another perspective, the star consists of six defects in the hexatic order, each with topological charge of +1/6. Five of these defects are at the ends of the arms, and one is at the center. The defects are connected by the five arms, which are domains walls in the tilt order. This structure was observed experimentally and analyzed theoretically in 1986, and it made a great impression on researchers at that time.

[4] S. B. Dierker, R. Pindak, and R. B. Meyer, "Consequences of Bond-Orientational Order on the Macroscopic Orientation Patterns of Thin Tilted Hexatic Liquid-Crystal Films," *Phys. Rev. Lett.* **56**, 1819 (1986).

7.3 Unequal Elastic Constants

So far, we have assumed that the free energy for a 2D polar or nematic phase is given by Eq. (7.1). However, that is not the most general form for the free energy. In general, there are different types of director gradients, and each type may have its own elastic constant. The distinction between different types of director gradients is very important in liquid crystals. In principle, it should also apply to magnets with spin-orbit coupling. However, this distinction is not often discussed in the literature on magnetism. Perhaps the difference between elastic constants might be smaller in magnets than in liquid crystals. That is why we consider this issue here, in a chapter about nematic order.

For a 2D director field $\hat{\boldsymbol{n}}(\boldsymbol{r})$ that depends on a 2D position $\boldsymbol{r}$, the director gradient tensor can be decomposed into two distinct elastic modes as[5]

$$\partial_i n_j = -n_i b_j + s(\delta_{ij} - n_i n_j). \tag{7.15}$$

Here, $\boldsymbol{b} = -(\hat{\boldsymbol{n}} \cdot \nabla)\hat{\boldsymbol{n}} = \hat{\boldsymbol{n}} \times (\nabla \times \hat{\boldsymbol{n}})$ is the bend vector, which shows how $\hat{\boldsymbol{n}}$ changes as we move parallel to $\hat{\boldsymbol{n}}$. For example, Fig. 7.2d shows a deformation with pure bend. By comparison, $s = \nabla \cdot \hat{\boldsymbol{n}}$ is the splay scalar, which shows how $\hat{\boldsymbol{n}}$ changes as we move perpendicular to $\hat{\boldsymbol{n}}$. The splay vector can be constructed as $\boldsymbol{s} = s\hat{\boldsymbol{n}}$. Figure 7.2c shows a deformation with pure 2D splay.

There is no symmetry reason why splay and bend need to have the same free energy cost. Rather, their elastic constants may be different. Hence, the 2D free energy can be written more generally as

$$F = \int d\boldsymbol{r} \left[\frac{1}{2}K_{11}s^2 + \frac{1}{2}K_{33}|\boldsymbol{b}|^2\right] = \int d\boldsymbol{r} \left[\frac{1}{2}K_{11}\left(\nabla \cdot \hat{\boldsymbol{n}}\right)^2 + \frac{1}{2}K_{33}\left|\nabla \times \hat{\boldsymbol{n}}\right|^2\right]. \tag{7.16}$$

This expression is the 2D version of the Oseen-Frank free energy for nematic liquid crystals. The coefficients K_{11} and K_{33} are elastic constants for splay and bend, respectively, in standard notation.[6] This free energy reduces to the simpler form of Eq. (7.1) in the special case of $K_{11} = K_{33} \equiv K$. Hence, that simpler form is often called the *single-elastic-constant approximation*. Another term for the simpler form with $K_{11} = K_{33}$ is *isotropic elasticity*, while the more general form with $K_{11} \neq K_{33}$ is *anisotropic elasticity*.

The difference between elastic constants is important for many areas of liquid crystal science. For topological defects, it has two main consequences:

---

[5] We should emphasize that a 2D director field is different from the more familiar case of a 3D director field. A 2D director field has two elastic modes, while a 3D director field has four elastic modes. See J. V. Selinger, "Interpretation of Saddle-Splay and the Oseen-Frank Free Energy in Liquid Crystals," *Liq. Cryst. Rev.* **6**, 129 (2018), Appendix 2.

[6] In this notation, K_{22} is the elastic constant for twist, which only occurs for 3D liquid crystals.

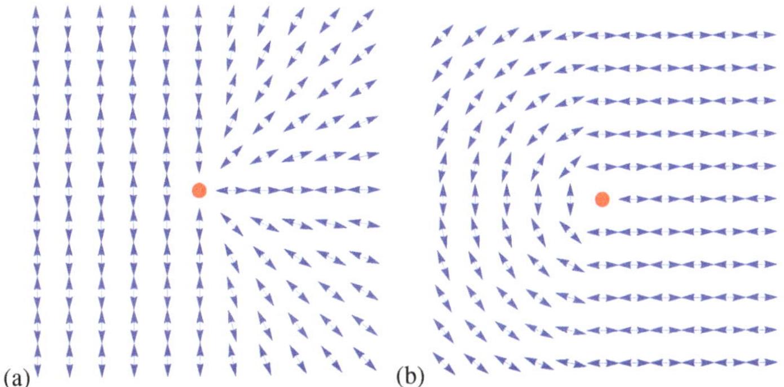

Fig. 7.10 Structure of a $+1/2$ defect in the limiting cases of extremely different elastic constants. (**a**) $K_{11} \ll K_{33}$. (**b**) $K_{11} \gg K_{33}$

First, for defects of topological charge $q = +1$, the two structures in Fig. 7.2c and d no longer have the same free energy. The defect in Fig. 7.2c has pure 2D splay, and hence its free energy is $F = \pi K_{11} \log(R/r_{\text{core}})$. By comparison, the defect in Fig. 7.2d has pure bend, so its free energy is $F = \pi K_{33} \log(R/r_{\text{core}})$. Any $+1$ defect will form with the structure of lower free energy, not the structure of higher free energy. Hence, $+1$ defects will look like Fig. 7.2c if $K_{11} < K_{33}$, or like Fig. 7.2d if $K_{11} > K_{33}$.

Second, for defects of other topological charges, there will generally be some combination of splay and bend. The defect can adjust its structure, so that it will have more of the lower-free-energy mode, and less of the higher-free-energy mode. Hence, the director angle around the defect will no longer follow $\theta = q\phi + \theta_0$ in polar coordinates (r, ϕ). To determine the optimum profile $\theta(\phi)$, we must solve the Euler-Lagrange equation associated with the free energy (7.16). In general, it cannot be solved exactly, but only numerically. As an illustration, Fig. 7.10 shows the structure of $+1/2$ defects in the limiting cases where the elastic constants K_{11} and K_{33} are very different. We can see that the director configurations adjusts to have all splay and no bend for $K_{11} \ll K_{33}$, or all bend and no splay for $K_{11} \gg K_{33}$. For other topological charges, the defect cannot completely eliminate the higher-energy mode, but it can reduce the amount of that mode.

7.4 Active Liquid Crystals

Over the past 10–20 years, *active matter* has become a large and growing area of research. The concept of active matter refers to systems that consume energy and convert it into large-scale, organized motion.[7] It was originally inspired by

[7] For a review of this research area, see M. C. Marchetti, J. F. Joanny, S. Ramaswamy, T. B. Liverpool, J. Prost, M. Rao, and R. A. Simha, "Hydrodynamics of Soft Active Matter," *Rev. Mod. Phys.* **85**, 1143 (2013).

the collective behavior of animals, such as flocks of birds, schools of fish, and herds of land animals, which spontaneously assemble and move together. These groups of animals exhibit orientational order, similar to ferromagnets or liquid crystals. Further research has found similar types of collective organization on the laboratory scale. For example, some experiments study bundles of microtubules driven by kinesin molecular motors, which use chemical energy from ATP to push the microtubules along their length. These systems form liquid-crystal-like phases that are constantly in motion, with topological defects forming and annihilating.[8] Other experiments put swimming bacteria into nematic liquid crystals, so that the orientation of the liquid crystal is coupled with the motion of the bacteria.[9]

Systems of active matter can have different types of order—nematic, polar, or even crystalline. They can be 2D or 3D. Very commonly, experimental studies of active matter involve 2D nematic order. That is why we discuss the topic of active matter in this chapter.

The fundamental feature of active systems is that they are not in thermal equilibrium. For that reason, the theory of active systems cannot be based on minimizing any free energy. Rather, it must be a dynamic theory to describe the collective motion. This motion is driven by the consumption of energy, and the direction of motion is controlled by the local orientational order. For one class of materials, called *extensile* materials, the motion is outward along the director $\hat{\boldsymbol{n}}$, inward perpendicular to $\hat{\boldsymbol{n}}$. For another class of materials, called *contractile* materials, the motion is inward along $\hat{\boldsymbol{n}}$, outward perpendicular to $\hat{\boldsymbol{n}}$.

Because of the collective motion, active systems generate many topological defects, always in pairs with positive and negative topological charges. Figure 7.11 illustrates the mechanism for defect formation. Here, the experimental images show an active nematic system of microtubule bundles driven by kinesin motors, which is *extensile*. Initially, the microtubule bundles have nematic domains with approximately uniform alignment. Because of the active force, the bundles extend along their length. This extension leads to a buckling instability, in which the material flows to the left, generating a *bend* of the director field. In the second image, the material flows further to the left, generating even more bend. When the bend becomes sharp enough, the material fractures along a horizontal line. The two ends of the line are $+1/2$ and $-1/2$ defects in the nematic order. In the third and fourth images, the material self-heals along the fracture line, and leaves behind the defects, which are now unbound. This process occurs throughout the system, leading to a proliferation of $+1/2$ and $-1/2$ defects.

[8] T. Sanchez, D. T. N. Chen, S. J. DeCamp, M. Heymann, and Z. Dogic, "Spontaneous Motion in Hierarchically Assembled Active Matter," *Nature* **491**, 431 (2012). In an ingenious illustration, this research group used an experimental video as the background for Vincent van Gogh's famous painting *Starry Night*; see https://www.youtube.com/watch?v=gB_LO-ODgkw.

[9] S. Zhou, A. Sokolov, O. D. Lavrentovich, and I. S. Aranson, "Living Liquid Crystals," *Proc. Natl. Acad. Sci. U.S.A.* **111**, 1265 (2014).

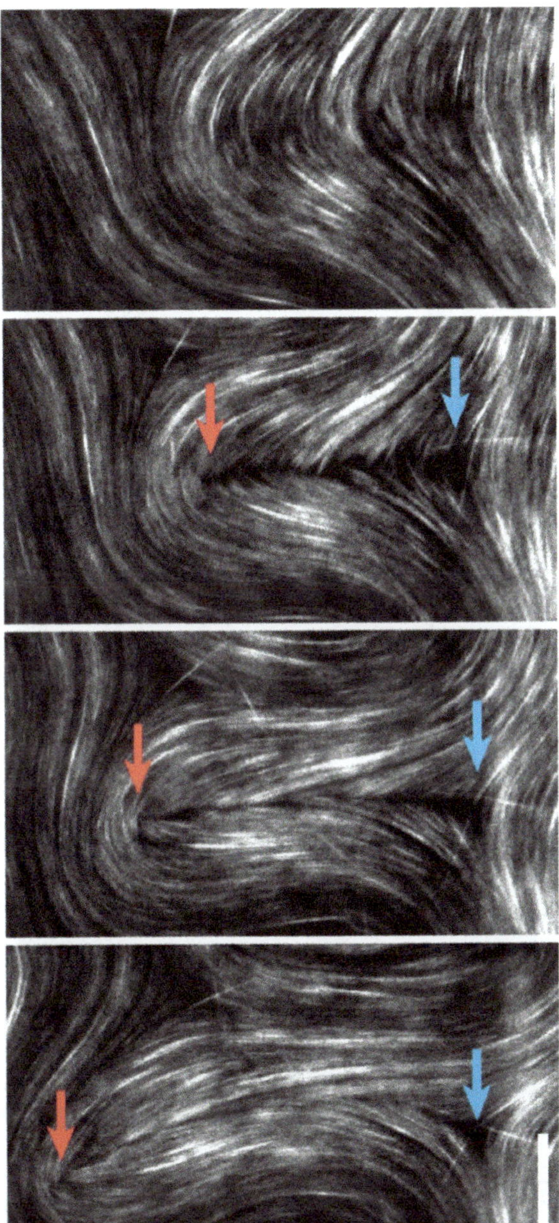

Fig. 7.11 Formation of a pair of topological defects in an active nematic system based on microtubules driven by kinesin. Time lapse is 15 s, and scale bar is 20 μm. Red and blue arrows indicate $+1/2$ and $-1/2$ defects, respectively. From T. Sanchez, D. T. N. Chen, S. J. DeCamp, M. Heymann, and Z. Dogic, *Nature* **491**, 431 (2012), reproduced with permission from SNCSC

After the defect pairs form, the $+1/2$ defects move rapidly. The direction of motion is determined by the defect orientation vector $\hat{\boldsymbol{p}}$, which was discussed in Sect. 7.1.4. In an extensile material, the $+1/2$ defects move in the $-\hat{\boldsymbol{p}}$ direction (comet heads forward). In a contractile material, they move in the $+\hat{\boldsymbol{p}}$ direction (comet tails forward). The motion can be modeled by hydrodynamic theories, which combine the orientational order and fluid flow.[10] This motion is one reason why the concept of defect orientation is more often studied for active than for conventional liquid crystals. By comparison, the $-1/2$ defects do not move very much. Because each $-1/2$ defect has a triad of three $\hat{\boldsymbol{p}}$ vectors, the active forces on the defect cancel each other.

While some defects are forming in pairs, other defects are annihilating in pairs. Eventually the system reaches a steady state, with a statistical average population of defects. In the steady state, the material is always in motion, swirling and undulating, and the $+1/2$ defects are all moving. The steady state is *not* thermal equilibrium, because it is continually consuming energy.

There is some evidence that the steady state can have orientational order of the $\hat{\boldsymbol{p}}$ orientation vectors for the defects themselves.[11] That orientational order is another reason why the concept of defect orientation is more often studied for active than for conventional liquid crystals.

---

[10] There have been many theoretical studies of the hydrodynamics of active liquid crystals. For three examples that emphasize the importance of the defect orientation vector, see S. Shankar, S. Ramaswamy, M. C. Marchetti, and M. J. Bowick, "Defect Unbinding in Active Nematics," *Phys. Rev. Lett.* **121**, 108002 (2018); X. Tang and J. V. Selinger, "Theory of Defect Motion in 2D Passive and Active Nematic Liquid Crystals," *Soft Matter* **15**, 587 (2019); S. Shankar and M. C. Marchetti, "Hydrodynamics of Active Defects: From Order to Chaos to Defect Ordering," *Phys. Rev. X* **9**, 041047 (2019).

[11] S. J. DeCamp, G. S. Redner, A. Baskaran, M. F. Hagan, and Z. Dogic, "Orientational Order of Motile Defects in Active Nematics," *Nat. Mater.* **14**, 1110 (2015).

3D Polar or Nematic Order

8

So far in this book, we have considered systems with 2D polar or nematic order. In this chapter, we will generalize the theory to consider 3D polar or nematic order. We will see that the 3D cases are quite different from the 2D cases, and that 3D polar order is quite different from 3D nematic order.

When we write about 3D order, we mean that the orientational order can point in any direction in 3D space. For that reason, it can be represented by a 3-component vector field $\hat{\boldsymbol{n}}(\boldsymbol{r})$, with

$$\hat{\boldsymbol{n}}(\boldsymbol{r}) = (n_x(\boldsymbol{r}), n_y(\boldsymbol{r}), n_z(\boldsymbol{r})). \tag{8.1}$$

As in the previous chapters, $\hat{\boldsymbol{n}}(\boldsymbol{r})$ describes the direction of orientational order, not the magnitude, and hence it is a unit vector field. In 3D, a unit vector has

$$|\hat{\boldsymbol{n}}|^2 = n_x^2 + n_y^2 + n_z^2 = 1. \tag{8.2}$$

We must distinguish carefully between 3D polar and nematic order:

- *Polar* means that $\hat{\boldsymbol{n}}$ and $-\hat{\boldsymbol{n}}$ represent two distinct physical states. In this case, the orientational order can be drawn as a single-headed arrow, as in Fig. 7.1a. This situation commonly occurs in magnetism. In the magnetic literature, it is sometimes called the *Heisenberg model*, as opposed to the xy model.
- *Nematic* means that $\hat{\boldsymbol{n}}$ and $-\hat{\boldsymbol{n}}$ represent the same physical state. For that reason, $\hat{\boldsymbol{n}}$ can be drawn as a double-headed arrow, as in Fig. 7.1b. This situation commonly occurs in 3D nematic liquid crystals. However, as mentioned in Sect. 7.1, there are exceptions to this general rule. Certain types of liquid crystals

J. V. Selinger, *Introduction to Topological Defects and Solitons*, Lecture Notes in Physics 1032, https://doi.org/10.1007/978-3-031-70200-6_8

can have polar order, such as nematic liquid crystals doped with magnetic platelets,[1] or the recently discovered ferroelectric nematic liquid crystals.[2]

We will discuss first the 3D polar case, and then the 3D nematic case.

8.1 No Disclinations with 3D Polar Order

In the previous chapters, we have seen that a system with 2D polar order can have topological defects. These defects are characterized by drawing a loop, and counting the winding number as we move around the loop. Orientational defects defined by this loop construction can be called *disclinations*.

For comparison, suppose that we have a system with 3D polar order. The question is: Can it have topological defects? To be more specific: Using the loop construction, can we distinguish between configurations of $\hat{\boldsymbol{n}}(\boldsymbol{r})$ with a defect vs. configurations of $\hat{\boldsymbol{n}}(\boldsymbol{r})$ without a defect?

To answer this question, we use a branch of mathematics called *homotopy theory*. The basic idea of homotopy theory is to study the relationship between position $\boldsymbol{r}$ and director $\hat{\boldsymbol{n}}$. To illustrate this relationship, we can draw a pair of two diagrams:

1. The first diagram shows *position space:* Just an ordinary picture of the system.
2. The second diagram shows *director space:* The set of all possible directors. It is also known as the *ground state manifold*, because each point in this set represents a possible ground state of the system.

The structure of director space depends on the dimensionality of $\hat{\boldsymbol{n}}$. Let us compare 2D and 3D:

For 2D polar order, the director $\hat{\boldsymbol{n}}$ must be a 2D unit vector with $|\hat{\boldsymbol{n}}|^2 = n_x^2 + n_y^2 = 1$. Hence, the set of all possible directors is the *unit circle*. In Fig. 8.1, we draw several examples of possible relationships between $\boldsymbol{r}$ and $\hat{\boldsymbol{n}}$. In each row, the left column shows position space, and the right column shows director space. The letters show the relationship between these spaces: position A corresponds to director A, position B corresponds to director B, and so forth. Suppose we move continuously along a closed loop in position space, from position A to B, C, D, E, F, and eventually back to A, as indicated by the green paths in the left column. Each loop corresponds to a loop in director space, from director A to B, C, D, E, F, and eventually back to A, as indicated by the green paths in the right column.

---

[1] A. Mertelj, D. Lisjak, M. Drofenik, and M. Čopič, "Ferromagnetism in Suspensions of Magnetic Platelets in Liquid Crystal," *Nature* **504**, 237 (2013).

[2] X. Chen, E. Korblova, D. Dong, X. Wei, R. Shao, L. Radzihovsky, M. A. Glaser, J. E. Maclennan, D. Bedrov, D. M. Walba, and N. A. Clark, "First-Principles Experimental Demonstration of Ferroelectricity in a Thermotropic Nematic Liquid Crystal: Polar Domains and Striking Electro-Optics," *Proc. Natl. Acad. Sci. U.S.A.* **117**, 14021 (2020).

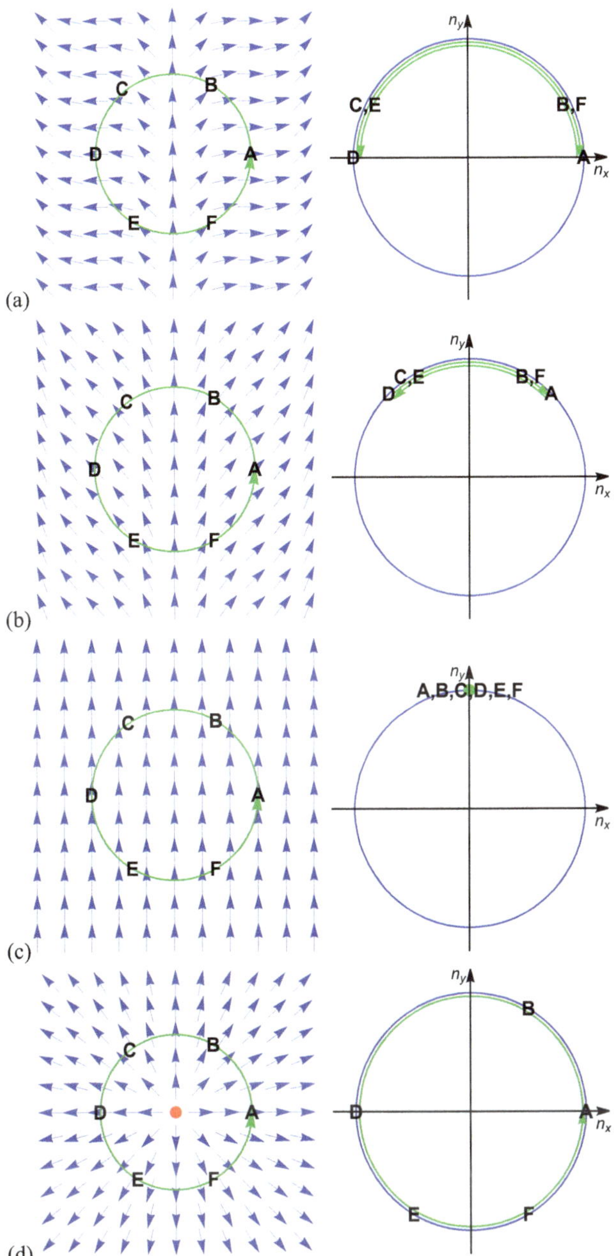

Fig. 8.1 Relationship between position space (left column) and director space (right column) for 2D polar order. In rows (**a**–**c**), the green path in position space does not enclose a topological defect. The corresponding green path in director space can be continuously shrunk down to a single point. In row (**d**), the green path in position space encloses a defect of winding number $+1$. The corresponding green path in director space goes all the way around the unit circle, and cannot be continuously shrunk down to a single point

In row (a), the green path in position space does not enclose a topological defect. The corresponding green path in director space goes part way around the unit circle, from A to D, and then it turns around and comes back to A. An important topological feature is: *This path in director space can be continuously shrunk down to a single point.* The shrinking process is shown explicitly in rows (b) and (c). From row (a) to (b), we slightly modify the structure in position space, and the path in director space becomes smaller. From row (b) to (c), the structure in position space becomes uniform, and the path in director space becomes a single point. In that sense, we would say that the structures in rows (a), (b), and (c) are topologically equivalent to each other. Indeed, any defect-free structure is topologically equivalent to a uniform structure.

By contrast, in row (d), the green path encloses a defect with winding number $+1$. The corresponding green path in director space goes all the way around the unit circle, exactly once, moving counter-clockwise. *This path in director space cannot be continuously shrunk down to a single point.* If we make any small changes to the green loop or to the structure in position space, the path in director space will still go all the way around the unit circle. This is a way of seeing that the winding number of the xy model is topologically protected, as already discussed in Chap. 1. Hence, for 2D polar order, the loop construction gives a clear topological distinction between structures with different winding numbers. It allows us to distinguish between structures with or without a defect.

Now, we can repeat the same type of analysis for the 3D case. For 3D polar order, the director $\hat{\boldsymbol{n}}$ must be a 3D unit vector with $|\hat{\boldsymbol{n}}|^2 = n_x^2 + n_y^2 + n_z^2 = 1$. Hence, the set of all possible directors is the *unit sphere*. Figure 8.2 shows several examples of possible relationships between $\boldsymbol{r}$ and $\hat{\boldsymbol{n}}$. In each row, position space is shown in the left column, and director space is shown in the right column. The green path through position space corresponds to the green path through director space.

Consider the director configuration in row (a). In position space, it might appear as if the green loop encloses a defect with winding number $+1$. In director space, the corresponding green loop goes around the equator of the unit sphere, moving from west to east (counter-clockwise when seen from the north). *However*, let us see what happens when we make continuous changes to the director configuration. As we go from row (a) to (b), the director gradually rotates upward at all positions. In director space, the corresponding green loop shifts upward, so that it is at a higher latitude around the unit sphere. From row (b) to (c), the director rotates upward even more, so that it is vertical and uniform at all positions. In the director space, the green loop shifts upward even more, so that it becomes a single point at the north pole. Hence, the structures in rows (a), (b), and (c) are topologically equivalent to each other. Because row (a) is topologically equivalent to a uniform structure, it is not really a topological defect after all.

By making further continuous changes in the director configuration, we can go from row (c) to (d) to (e). In director space, the green loop moves downward in latitude, now going from east to west (clockwise when seen from the north). In row (e), the green loop is back on the equator around the unit sphere. In position space, the corresponding loop might appear to enclose a defect with winding number -1.

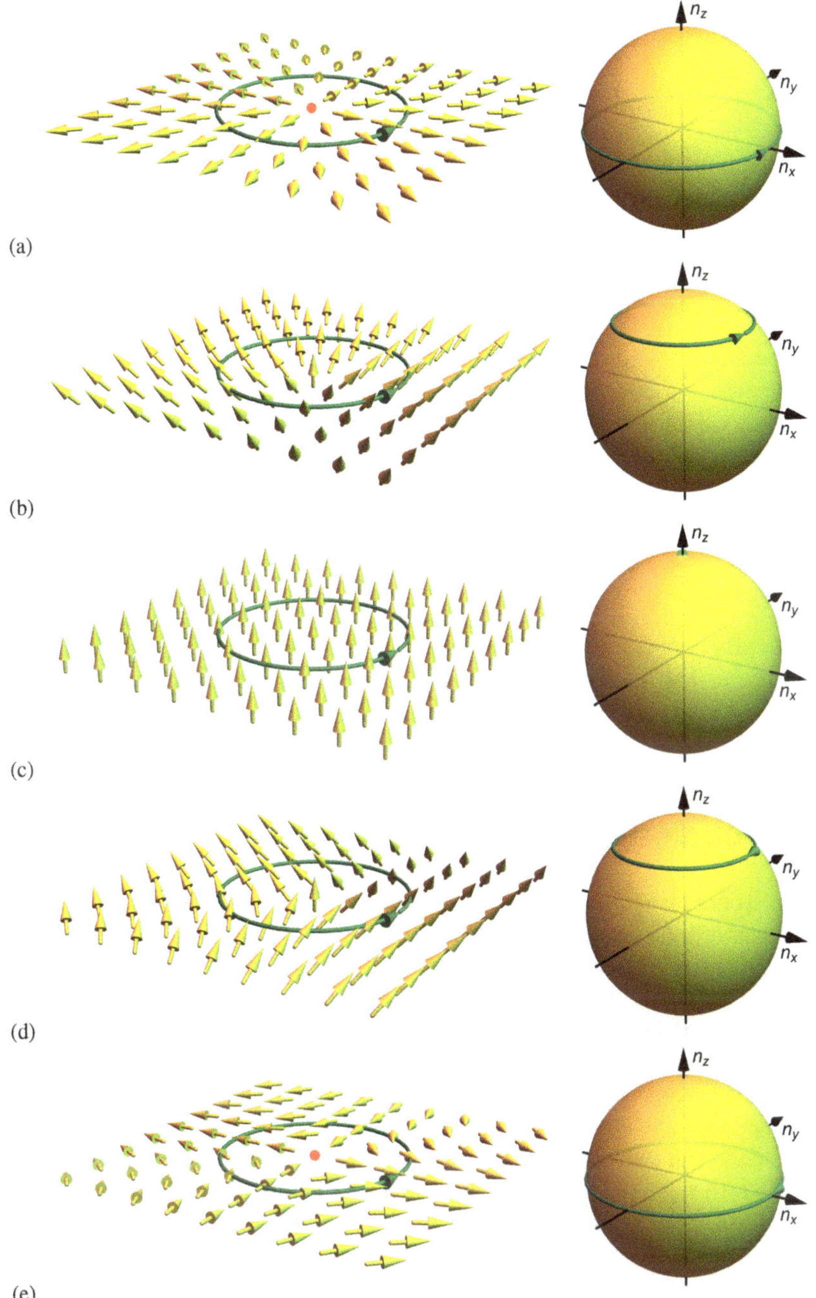

Fig. 8.2 Relationship between position space (left column) and director space (right column) for 3D polar order. As discussed in the text, we can continuously transform from any row to any other row in this figure. Hence, all of these director configurations are equivalent to the uniform structure in row (**c**), and none of them is a topological defect

From this continuous transformation, we can conclude that row (e) is topologically equivalent to a uniform structure, and hence it is not really a topological defect either.

For 3D polar order, the same argument applies to apparent defects with *any* winding number. They can all be continuously transformed into uniform director configurations, and hence they are not really topological defects.

This continuous transformation of an apparent defect into a uniform director configuration is often called *escape into the third dimension*, because the director uses the third dimension to avoid the need for a 2D defect. It is also described by the phrase: "*You can't lasso a sphere.*" A lasso is a loop of rope used by a cowboy. This phrase suggests that a cowboy might *try* to capture a sphere by throwing a loop of rope around it, like the green loop in Fig. 8.2a or e. However, the rope would slide upwards or downwards, and slip off of the sphere, as in Fig. 8.2c. By contrast, a cowboy *can* lasso a circle, as in Fig. 8.1d, because the rope goes around the circle and cannot slip off.

From the argument in this section, we see that there is a profound difference between 2D polar order and 3D polar order. This difference occurs because of the shape of director space: the unit circle compared with the unit sphere. For 2D polar order, there are distinct director configurations with winding number of $\ldots, -2, -1, 0, +1, +2, \ldots$, and it is impossible to continuously transform one into another. By contrast, for 3D polar order, these configurations can be continuously transformed into each other, and none of them is a topological defect. Thus, a system with 3D polar order does not have any disclinations—i.e., topological defects characterized by a winding number, defined through the loop construction.[3]

8.2 Disclinations with 3D Nematic Order

In this section, we repeat the analysis of the previous section, but for *nematic* rather than polar order. Once again, we consider first the 2D case and then the 3D case.

For 2D nematic order, the director $\hat{\boldsymbol{n}}$ is a 2D unit vector, and we understand that $\hat{\boldsymbol{n}}$ and $-\hat{\boldsymbol{n}}$ represent the same physical state. For director space, we do not want to use the unit circle, because it counts each physical state twice—once with $\hat{\boldsymbol{n}}$ and once with $-\hat{\boldsymbol{n}}$. Instead, we should really use a *unit semicircle*, which counts each physical state exactly once. It does not matter whether we use the unit semicircle with $n_x \geq 0$, or with $n_y \geq 0$, or any other half of the circle. To be specific, we will use the unit semicircle with $n_y \geq 0$. We must remember that $\hat{\boldsymbol{n}} = (+1, 0)$ and $\hat{\boldsymbol{n}} = (-1, 0)$ represent the same physical state, so those two ends of the semicircle are equivalent.

[3] In Chap. 10, we will see that there are other types of topological defects, not characterized by a winding number defined through the loop construction. A system with 3D polar order may still have those defects.

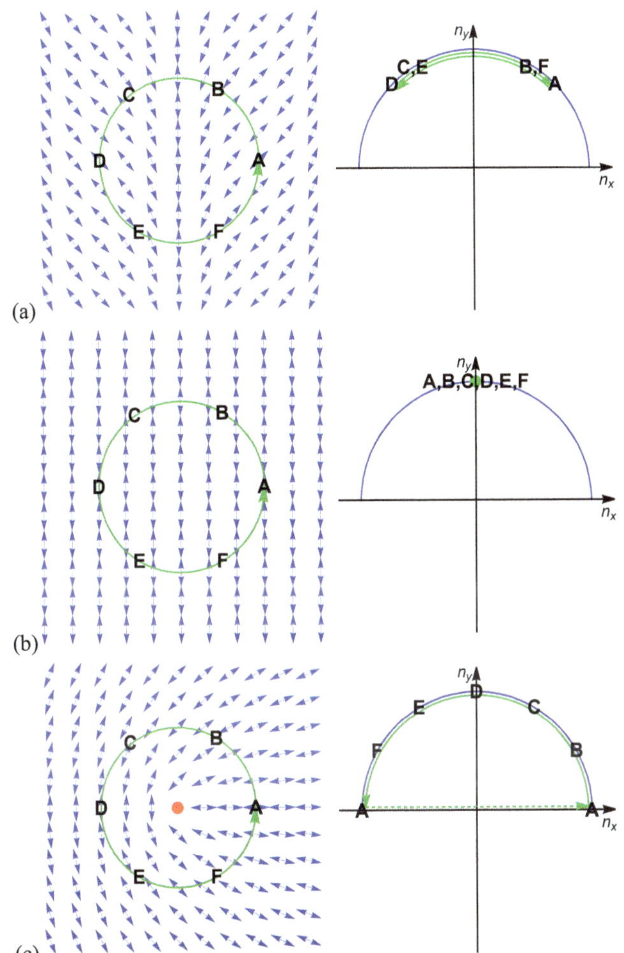

Fig. 8.3 Position space (left column) and director space (right column) for 2D nematic order. In rows (**a–b**), the green loop in position space does not enclose a topological defect, and the corresponding loop in director space can be shrunk down to a single point. In row (**c**), the green loop in position space encloses a defect of winding number $+1/2$. The corresponding loop in director space goes all the way around the semicircle and returns to the beginning (because the two points marked A represent the same physical state), so it cannot be shrunk down to a single point

The left column of Fig. 8.3 shows some possible director configurations in position space, with $\hat{\boldsymbol{n}}(\boldsymbol{r})$ indicated by blue double-headed arrows. In each case, we have drawn a green loop from position A to B, C, D, E, F, and back to A. The right column shows the corresponding loop on the unit semicircle in director space.

Our discussion of these configurations is very similar to our discussion for the 2D polar case. In row (a), the green loop in position space does not enclose a topological defect. The corresponding loop in director space goes part way around the unit semicircle, and then turns around and comes back the starting point. This path in director space can be continuously shrunk down to a single point. We can see this shrinking process explicitly as we go from row (a) to (b), gradually modifying the director configuration.

In row (c), the green loop in position space encloses a topological defect of winding number $+1/2$. The corresponding loop in director space goes around the unit semicircle exactly once, moving counter-clockwise. When it reaches the endpoint of the semicircle, it is back to the beginning, because the two points marked A represent the same physical state. This path in director space cannot be continuously shrunk down to a single point. If we make any small changes to the green loop or to the structure in position space, the path in director space must still go all the way around the unit semicircle, using the equivalence between the two opposite points. Hence, just as in the 2D polar case, the loop construction gives a clear topological distinction between structures with different winding numbers, with or without a defect.

Finally, let us do this analysis for 3D nematic order. In this case, the director $\hat{\boldsymbol{n}}$ is a 3D unit vector, and we still understand that $\hat{\boldsymbol{n}}$ and $-\hat{\boldsymbol{n}}$ represent the same physical state. For director space, we should not use the unit sphere, because it would count each physical state twice. Instead, we should use a *unit hemisphere*. It does not matter which half of the sphere we use. For convenience, we will choose the northern hemisphere, with $n_z \geq 0$, as shown in the right column of Fig. 8.4. This hemisphere is bordered by the equator, where $n_z = 0$. On the equator, there is still a slight problem of double counting, so we must remember that the points $(n_x, n_y, 0)$ and $(-n_x, -n_y, 0)$ are equivalent. By analogy, on a globe of the Earth, Singapore is equivalent to Quito, Ecuador, because they are at opposite points on the Earth's equator.

The left column of Fig. 8.4 shows some director configurations in position space. Instead of double-headed arrows, we have represented $\hat{\boldsymbol{n}}(\boldsymbol{r})$ by yellow cylinders, which are zero-headed arrows, just because they are easier to visualize. The right column shows the unit hemisphere for director space.

In row (a), the green loop in position space encloses a topological defect of winding number $+1/2$. The corresponding loop in director space begins at the point $\hat{\boldsymbol{n}} = (1, 0, 0)$ and goes halfway around the equator, moving from west to east (counter-clockwise seen from the north). When it reaches the point $\hat{\boldsymbol{n}} = (-1, 0, 0)$, it is back at the beginning, because these two points represent the same physical state. Hence, the green path in director space is a closed loop. The important topological feature is: *This path in director space cannot be continuously shrunk down to a single point.* It cannot slide up to the north pole, because (n_x, n_y, n_z) is not equivalent to $(-n_x, -n_y, +n_z)$ for $n_z > 0$. Thus, we have successfully lassoed the unit hemisphere, using the equivalence between opposite points on the equator. The director configuration in row (a) is *not* topologically equivalent to a uniform state. Rather, it is a topological defect, as identified by the loop construction.

If we rotate the entire director configuration about the x-axis, we continuously move through the series of structures shown in rows (a–e). The corresponding paths in director space do not remain on the equator. Rather, they go up to

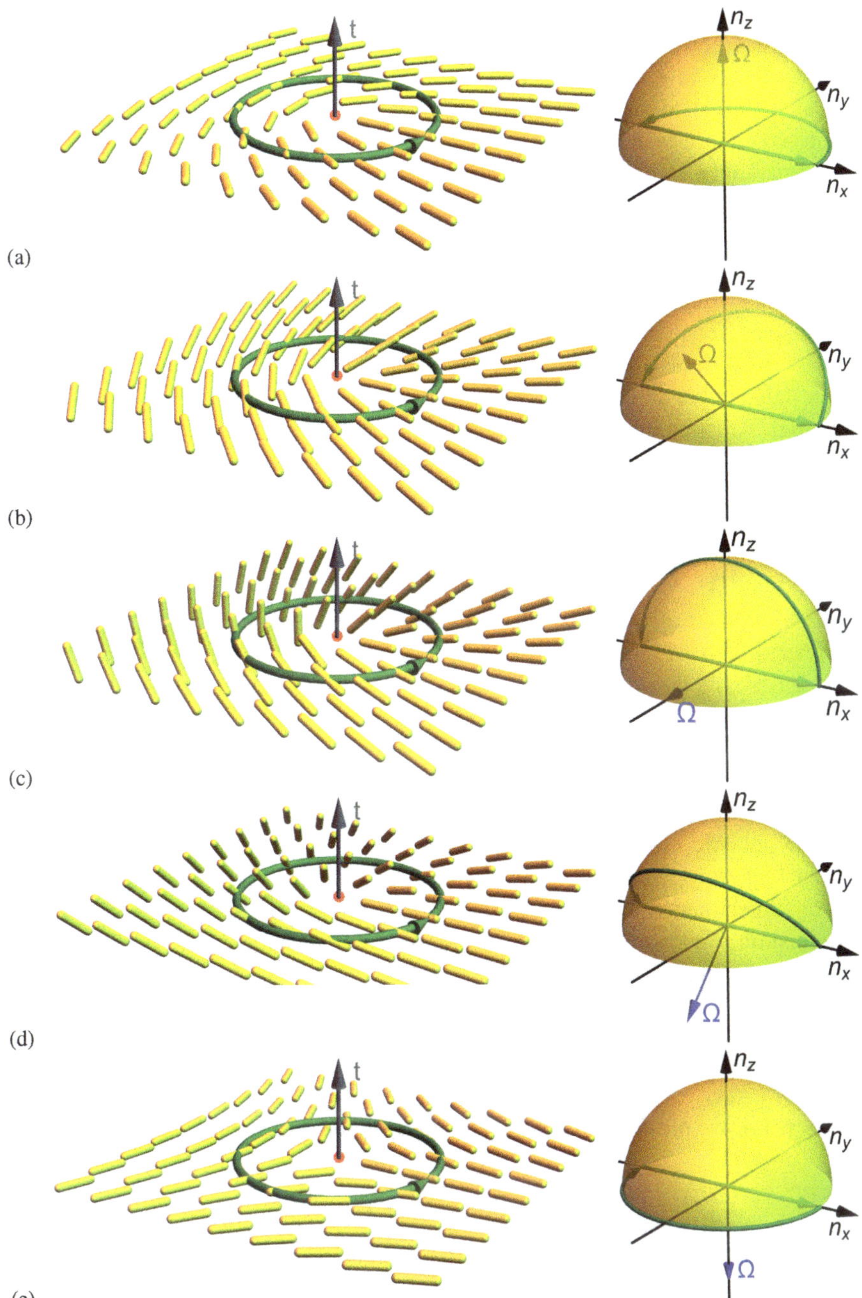

Fig. 8.4 Position space (left column) and director space (right column) for 3D nematic order. In all rows, the green loop encloses a half-charged disclination, indicated by a red dot. These disclinations can be continuously transformed into each other. The vectors $\hat{\boldsymbol{t}}$ and $\hat{\boldsymbol{\Omega}}$ are discussed in Sect. 8.4.2. The angle between these vectors is: **(a)** $\beta = 0$. **(b)** $\beta = \pi/4$. **(c)** $\beta = \pi/2$. **(d)** $\beta = 3\pi/4$. **(e)** $\beta = \pi$

higher latitudes and come back down. They are still closed loops, because of the equivalence between opposite points on the equator. Hence, all of these director configurations have topological defects, as identified by the loop construction. They are all topologically equivalent to each other, because they can be continuously transformed into each other.

Remarkably, in row (e), the green loop in position space encloses a topological defect of winding number $-1/2$. The corresponding loop in director space goes halfway around the equator, moving from east to west (clockwise seen from the north). The series of structures in Fig. 8.4 shows that a $+1/2$ defect in row (a) can be continuously transformed into a $-1/2$ defect in row (e) by rotating through the third dimension. Hence, for 3D nematic order, a $+1/2$ defect is topologically equivalent to a $-1/2$ defect. Likewise, any half-integer defect can be transformed into any other half-integer defect. Based on this argument, for 3D nematic order, there is no topological distinction between defects with different half-integer winding numbers. Rather, they are all just *half-charged disclinations*. However, there *is* a topological distinction between a half-charged disclination and a defect-free configuration.

What about integer-charged disclinations for 3D nematic order? To answer that question, we can re-use Fig. 8.2, but now we mentally erase the yellow arrowheads and only think about the yellow cylinders representing the director configuration. Even without the arrowheads, this figure shows that an apparent defect with winding number $+1$ or -1 can be continuously transformed into a uniform configuration, through escape into the third dimension. The same is true for an apparent defect with any integer winding number. Hence, for 3D nematic order, there are no integer-charge disclinations that are topologically distinct from the uniform configuration.

To summarize the results of this section: For 2D nematic order, there are many types of disclinations, with any integer or half-integer winding number. However, for 3D nematic order, there is only one type of disclination, which can just be called *half-charged*, without specifying whether it is $+1/2$ or $-1/2$. Any half-charged disclination can be continuously transformed into any other half-charged disclination, and any integer-charged apparent disclination can be continuously transformed into a defect-free state.

8.3 Topological Notation

In Sect. 1.4, we presented a topological notation for defects in 2D polar order:

$$\pi_1(S^1) = \mathbb{Z}. \tag{8.3}$$

As a reminder, the symbol π_1 means the first homotopy group, which describes the classification of defects based on 1D measuring surfaces—i.e., the loop construction to determine the winding number. The symbol S^1 represents the unit circle, which is a 1D version of a sphere. It is the director space for 2D polar order. The symbol $\mathbb{Z}$ refers to the set of all integers (positive, negative, and zero). Hence, Eq. (8.3) states that 2D polar order has disclinations that can be classified by a single integer, which is the winding number, defined through the loop construction.

We can now make analogous statements for the other forms of order discussed in this chapter.

For 2D nematic order, the director space is the unit semicircle, with equivalence between the two endpoints. This semicircle has the same topological properties as a full circle, because it is connected in a single loop. In that sense, it is similar to S^1. The defects can still be classified by a winding number, which is an integer or half-integer. That set of possible winding numbers is similar to $\mathbb{Z}$, because it is countable and goes to infinity in both directions. Hence, Eq. (8.3) still applies.

For 3D polar order, the director space is the unit sphere, which is written as S^2 because it is a 2D surface. We have seen that it does not have any topological defects characterized by a winding number, defined through the loop construction. Hence, the mathematical statement is

$$\pi_1(S^2) = 0, \tag{8.4}$$

where the 0 means no such defects.

For 3D nematic order, the director space is the unit hemisphere, with the understanding that opposite points on the equator are equivalent. This unit hemisphere is written as $S^2/\mathbb{Z}_2$. It is also called the *real projective plane*, written as $\mathbb{R}P^2$. As far as the loop construction is concerned, this system has two types of director configurations: either defect-free or half-charged disclination. That set of two possibilities is written as $\mathbb{Z}_2$. The mathematical statement then becomes

$$\pi_1(S^2/\mathbb{Z}_2) = \pi_1(\mathbb{R}P^2) = \mathbb{Z}_2. \tag{8.5}$$

All of these equations have the symbol π_1 because we are still working with 1D measuring surfaces. We will discuss higher-dimensional measuring surfaces in Chaps. 10 and 11.

8.4 Geometric and Physical Properties of Disclinations

We have seen that the topological classification of disclinations for 3D nematic order is very simple. A director configuration might be defect-free, or it might have a half-charged disclination. However, topology does not give a full description of disclinations. Rather, disclinations are physical objects with several geometric features, and these geometric features are important for the physics of disclinations. In this section, we discuss the geometric and physical properties.

8.4.1 Disclination Lines and Loops in 3D

So far, we have considered the configuration of $\hat{\boldsymbol{n}}(\boldsymbol{r})$ around a disclination, for position $\boldsymbol{r}$ in a 2D plane. In this plane, a disclination is a *point defect*, as illustrated by the red dots in the left column of Fig. 8.4.

We can now extend this concept to a full 3D system, where both $\hat{\boldsymbol{n}}$ and $\boldsymbol{r}$ are 3D vectors. For a 3D system, a disclination becomes a *line defect*. The generalization from a point defect in 2D to a line defect in 3D is quite analogous to our previous discussion for the xy model in Sect. 1.3.

A disclination line may be straight or curved. It cannot begin or end anywhere inside a system. It might go all the way to the boundary of a system, or it might curve into a closed loop. Furthermore, the director configuration around the disclination line might remain the same along the line, or the director configuration might vary as a function of position.

Figure 8.5 shows five possibilities for the structure of a closed, circular disclination loop. Before we discuss these possibilities, we should make a comment about the visualization. In these 3D systems, $\hat{\boldsymbol{n}}(\boldsymbol{r})$ is defined *everywhere*, except on the disclination loop itself. However, it is difficult to draw a readable picture of $\hat{\boldsymbol{n}}(\boldsymbol{r})$ everywhere in 3D space. For that reason, we only draw $\hat{\boldsymbol{n}}(\boldsymbol{r})$ as yellow cylinders at a few positions around the disclination loop, to show the configuration close to the loop itself. Alternative visualizations are shown in Fig. 10.13, or readers may make their own plots of the director fields

$$\hat{\boldsymbol{n}}_a(\boldsymbol{r}) = \hat{\boldsymbol{\rho}} \cos\theta + \hat{\boldsymbol{z}} \sin\theta, \tag{8.6a}$$

$$\hat{\boldsymbol{n}}_b(\boldsymbol{r}) = \hat{\boldsymbol{\phi}} \cos\theta + \hat{\boldsymbol{z}} \sin\theta, \tag{8.6b}$$

$$\hat{\boldsymbol{n}}_c(\boldsymbol{r}) = -\hat{\boldsymbol{\rho}} \cos\theta + \hat{\boldsymbol{z}} \sin\theta, \tag{8.6c}$$

$$\hat{\boldsymbol{n}}_d(\boldsymbol{r}) = \hat{\boldsymbol{y}} \cos\theta + \hat{\boldsymbol{z}} \sin\theta, \tag{8.6d}$$

$$\hat{\boldsymbol{n}}_e(\boldsymbol{r}) = \hat{\boldsymbol{y}} \cos\theta + \hat{\boldsymbol{x}} \sin\theta, \tag{8.6e}$$

where

$$x = \rho\cos\phi, \quad y = \rho\sin\phi, \quad \rho = \sqrt{x^2 + y^2}, \quad \phi = \tan^{-1}\left(\frac{y}{x}\right), \tag{8.7a}$$

$$\hat{\boldsymbol{\rho}} = \hat{\boldsymbol{x}} \cos\phi + \hat{\boldsymbol{y}} \sin\phi, \quad \hat{\boldsymbol{\phi}} = -\hat{\boldsymbol{x}} \sin\phi + \hat{\boldsymbol{y}} \cos\phi, \tag{8.7b}$$

$$\theta = \frac{1}{2}\tan^{-1}\left(\frac{z}{\rho - 1}\right) + \frac{1}{2}\tan^{-1}\left(\frac{z}{\rho + 1}\right). \tag{8.7c}$$

With that comment, let us proceed. In each part of Fig. 8.5, the red loop shows the disclination line. At four positions along the loop, we have drawn green measuring loops, so that we can determine whether the director configuration along the green loop rotates through a semicircle, and hence determine whether the green loop encloses a disclination. We must emphasize that the red loop is a physical object,

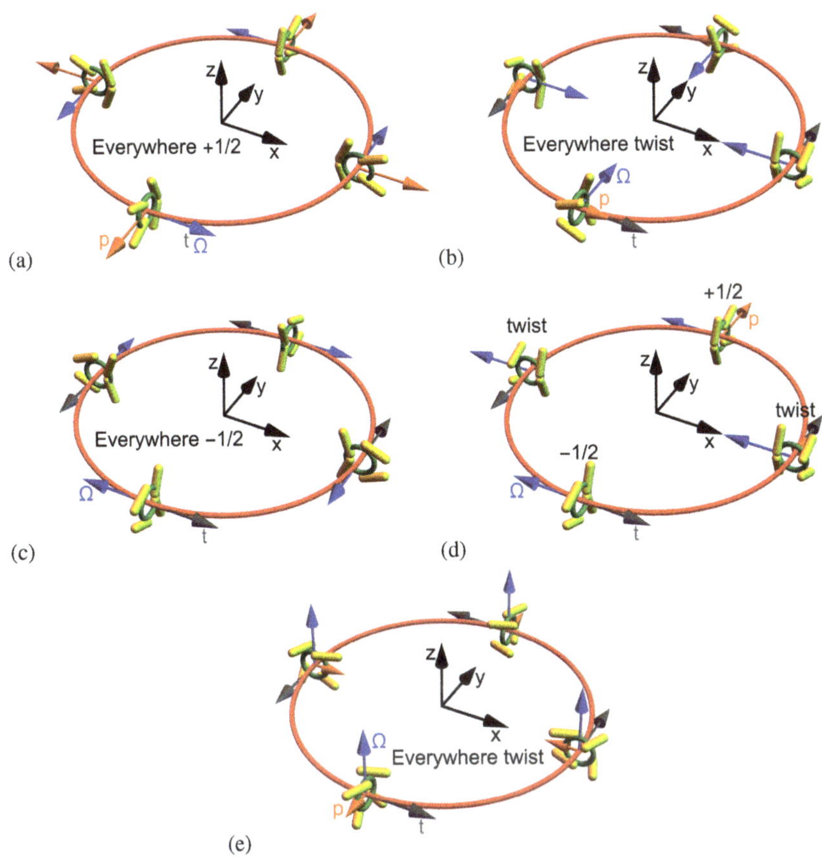

Fig. 8.5 Five examples for the geometric structure of a closed, circular disclination loop. (**a**) Everywhere $+1/2$ wedge. (**b**) Everywhere twist with $\hat{\boldsymbol{\Omega}}$ inward. (**c**) Everywhere $-1/2$ wedge. (**d**) Varying as a function of position along the loop. (**e**) Everywhere twist with $\hat{\boldsymbol{\Omega}}$ upward. In each part, the red loop is the physical disclination loop, and the green loops are just measuring loops to determine whether a disclination is present. The director $\hat{\boldsymbol{n}}(\boldsymbol{r})$ is defined everywhere (except right on the disclination), but it is only shown at a few points near the disclination. The tangent vector $\hat{\boldsymbol{t}}$ is indicated by gray arrows, the rotation vector $\hat{\boldsymbol{\Omega}}$ by blue arrows, and the orientation vector $\boldsymbol{p}$ by orange arrows. (In part (**a**), the gray arrows are covered by blue arrows.) Alternative visualizations are shown in Fig. 10.13

but the green loops are *not* physical objects; they are just mathematical constructions to determine whether a disclination is present there. On each green loop, we have drawn $\hat{\boldsymbol{n}}(\boldsymbol{r})$ at four positions, so that we can see the local director configuration near the disclination.

In Fig. 8.5a–c, the director configuration remains the same everywhere along the red disclination loop. These three pictures show three possible types of director configurations near a disclination:

1. In Fig. 8.5a, the director configuration lies in the plane perpendicular to the disclination line, and it looks like a $+1/2$ defect, as in the left column of Fig. 8.4a. It is called a "$+1/2$ *wedge disclination*."
2. In Fig. 8.5b, the director configuration has a 3D twisted structure, as in the left column of Fig. 8.4c. It is called a "*twist disclination*."
3. In Fig. 8.5c, the director configuration lies in the plane perpendicular to the disclination line, and it looks like a $-1/2$ structure, as in the left column of Fig. 8.4e. It is called a "$-1/2$ *wedge disclination*."

By comparison, in Fig. 8.5d, the director configuration varies as a function of position along the red disclination loop. Along the $+y$-axis, the structure is a $+1/2$ wedge disclination. Along the $-x$-axis, it is a twist disclination. Along the $-y$-axis, it is a $-1/2$ wedge disclination. Along the $+x$-axis, it is a twist disclination again. Between these positions, the structure gradually changes between those configurations, as in Fig. 8.4b and d.

In Fig. 8.5e, the director configuration also varies as a function of position along the loop. However, the structure is always a twist disclination.

From these examples, we see that "$+1/2$ wedge," "twist," and "$-1/2$ wedge" are useful terms to describe the director configuration around a disclination. There is no sharp topological distinction between these cases, but the terms still provide information about the local geometry. In the following section, we will make these concepts more quantitative.

8.4.2 Tangent and Rotation Vectors, Other Orientational Properties

To characterize the local director configuration around a disclination, we can define several vectors and tensors. The simplest geometric feature to describe is the local orientation of the disclination line. This orientation can be represented by the tangent vector $\hat{\boldsymbol{t}}$, which is the unit vector tangent to the line at any particular point. This vector is indicated by the gray arrows in each part of Fig. 8.5. (In part (a), the gray arrows are covered by the blue arrows discussed below, and only the gray label is visible.)

Defining the tangent vector $\hat{\boldsymbol{t}}$ requires one arbitrary choice: We must decide which way is *forward* or *backward* along the disclination line. If we switch the choice of forward and backward, then $\hat{\boldsymbol{t}}$ changes to $-\hat{\boldsymbol{t}}$ everywhere along the line. Our description of the line must be invariant under this transformation.

The sign of $\hat{\boldsymbol{t}}$ is not related to the right-hand rule. Hence, $\hat{\boldsymbol{t}}$ is a proper vector, not a pseudovector.

Another important geometric feature is represented by the rotation vector $\hat{\boldsymbol{\Omega}}$.[4] As we move in a loop around the disclination in position space, the director rotates by π radians about the unit vector $\hat{\boldsymbol{\Omega}}$ in director space. In other words, the green semicircle in the right column of Fig. 8.4 defines a plane, and $\hat{\boldsymbol{\Omega}}$ is perpendicular to that plane. It is indicated by the blue arrows in each part of Fig. 8.5.

The sign of $\hat{\boldsymbol{\Omega}}$ is determined through the following procedure:

1. We make an arbitrary choice for the sign of $\hat{\boldsymbol{t}}$.
2. Based on the sign of $\hat{\boldsymbol{t}}$, the right-hand rule tells us which way to move around the measuring loop in position space.
3. Moving in a specific direction around the loop in position space corresponds to a specific direction around the loop in director space.
4. Based on the direction around the loop in director space, the right-hand rule tells us the sign of $\hat{\boldsymbol{\Omega}}$.

From this procedure, we can see that the sign of $\hat{\boldsymbol{\Omega}}$ involves the same arbitrary choice as the sign of $\hat{\boldsymbol{t}}$, so $\hat{\boldsymbol{\Omega}}$ changes sign if $\hat{\boldsymbol{t}}$ changes sign. The sign of $\hat{\boldsymbol{\Omega}}$ does not involve an arbitrary choice of right-hand rule vs. left-hand rule, because the procedure uses the right-hand rule twice. Hence, $\hat{\boldsymbol{\Omega}}$ is also a proper vector rather than a pseudovector.

We now consider the relationship between $\hat{\boldsymbol{t}}$ and $\hat{\boldsymbol{\Omega}}$. Let β be the angle between these vectors, defined by

$$\cos\beta = \hat{\boldsymbol{t}} \cdot \hat{\boldsymbol{\Omega}}. \tag{8.8}$$

This angle is invariant if $\hat{\boldsymbol{t}}$ and $\hat{\boldsymbol{\Omega}}$ both change sign.

The examples in Fig. 8.5 show that $\hat{\boldsymbol{t}}$ and $\hat{\boldsymbol{\Omega}}$ can change as we move along a disclination line. In parts (a–c), $\hat{\boldsymbol{t}}$ and $\hat{\boldsymbol{\Omega}}$ change together, keeping a constant angle β. In parts (d–e), $\hat{\boldsymbol{t}}$ changes while $\hat{\boldsymbol{\Omega}}$ remains constant. The angle β changes in part (d), but remains constant in part (e).

From the relationship between $\hat{\boldsymbol{t}}$ and $\hat{\boldsymbol{\Omega}}$, characterized by β, we can determine the local geometry of the disclination. Three cases are:

- If $\hat{\boldsymbol{t}}$ and $\hat{\boldsymbol{\Omega}}$ are *parallel*, so that $\beta = 0$, then the disclination has a $+1/2$ wedge structure.
- If $\hat{\boldsymbol{t}}$ and $\hat{\boldsymbol{\Omega}}$ are *perpendicular*, so that $\beta = \pi/2$, then the disclination has a twist structure.
- If $\hat{\boldsymbol{t}}$ and $\hat{\boldsymbol{\Omega}}$ are *antiparallel*, so that $\beta = \pi$, then the disclination has a $-1/2$ wedge structure.

[4] To our knowledge, the rotation vector was first defined by J. Friedel and P.-G. de Gennes, "Boucles de disclination dans les cristaux liquides," *C. R. Acad. Sc. Paris B* **268**, 257 (1969).

Intermediate values of β give intermediate types of geometric structure.[5]

The vectors $\hat{\boldsymbol{t}}$ and $\hat{\boldsymbol{\Omega}}$ do not completely characterize the local geometry of a disclination. In Sect. 7.1.4, we already showed that $\pm 1/2$ defects in 2D nematic order can be regarded as objects with an orientation. Now, for 3D nematic order, we see that a disclination line can have a local geometric structure similar to a $\pm 1/2$ defect. Hence, a disclination line must also have orientational features analogous to $\pm 1/2$ defects.

The orientational features of disclination lines have been studied in a recent paper.[6] This paper shows that the orientational features can be separated into three parts, based on how they transform under rotations about $\hat{\boldsymbol{\Omega}}$. The first part involves a vector, the second part involves a tensor of rank two, and the third part involves a tensor of rank three.

Here, let us consider the vector part. For 2D nematic order, a $+1/2$ defect has an orientation vector $\hat{\boldsymbol{p}}$, defined in Eq. (7.11), in the direction of $\nabla \cdot (\hat{\boldsymbol{n}}\hat{\boldsymbol{n}})$. For 3D nematic order, a disclination has an analogous orientation vector $\boldsymbol{p}$. In the 3D case, $\boldsymbol{p}$ is not a unit vector, but rather has magnitude $|\boldsymbol{p}| = \frac{1}{2}(1 + \cos\beta)$. Hence, it is largest for a $+1/2$ wedge disclination, has intermediate magnitude for a twist disclination, and vanishes for a $-1/2$ wedge disclination. This vector is indicated by the orange arrows in Fig. 8.5. We can see that these arrows are longest in part (a), intermediate in parts (b) and (e), vanish in part (c), and have varying length in part (d). In general, $\boldsymbol{p}$ represents the orientation of the $+1/2$ aspect of disclination geometry, with onefold rotational symmetry. It is especially important for *active* 3D nematic liquid crystals, because it gives the direction of the local active force on the disclination.

For the part involving a tensor of rank two, the magnitude is $\sin\beta$. This part is largest for a twist disclination, and vanishes for a $\pm 1/2$ wedge disclination. It represents the orientation of the twist aspect of disclination geometry, with twofold rotational symmetry. For the part involving a tensor of rank three, the magnitude is $\frac{1}{2}(1 - \cos\beta)$. It is largest for a $-1/2$ wedge disclination, intermediate for a twist disclination, and zero for a $+1/2$ wedge disclination. It represents the orientation of the $-1/2$ wedge aspect of disclination geometry, with threefold rotational symmetry. These two parts are not illustrated here.

[5] As an aside, we might mention one *hypothetical* situation: Suppose that $\hat{\boldsymbol{n}}$ were defined in some quantum space, like a higher-dimensional version of a superfluid phase angle. In that case, $\hat{\boldsymbol{\Omega}}$ would also be defined in that quantum space, while $\hat{\boldsymbol{t}}$ would still be defined in real space. The dot product $\hat{\boldsymbol{t}} \cdot \hat{\boldsymbol{\Omega}}$ would then be undefined, and the angle β would be undefined. Hence, this system would not have any geometric distinction between $+1/2$ wedge, twist, and $-1/2$ wedge disclinations. It would only have a topological distinction of half-charged disclinations vs. no disclinations. However, that hypothetical situation does *not* apply to real liquid crystals, because $\hat{\boldsymbol{n}}$ is a direction in real space.

[6] C. Long, X. Tang, R. L. B. Selinger, and J. V. Selinger, "Geometry and Mechanics of Disclination Lines in 3D Nematic Liquid Crystals," *Soft Matter* **17**, 2265 (2021).

8.4.3 Free Energy and Drag

Because a disclination line is a physical object, it has physical properties, including a free energy and a drag coefficient. In this section, we briefly discuss those properties.

A single, isolated disclination line has a free energy per length. This free energy per length can be interpreted as a line tension, because it shows how the total free energy of a system increases if the disclination line becomes longer. For a simple model, we can follow the same argument as in Sect. 3.1. We assume:

- The elastic free energy has the form
$$F = \frac{1}{2} K \int d\boldsymbol{r} |\nabla \hat{\boldsymbol{n}}|^2, \tag{8.9}$$
with only a single Frank elastic constant K.[7]
- The disclination line is straight, so that the tangent vector $\hat{\boldsymbol{t}}$ is constant along the line.
- The rotation vector $\hat{\boldsymbol{\Omega}}$ and other orientational properties are also constant along the line.

In that case, Eq. (3.10) gives the free energy per length, or line tension,

$$F_{\text{disclination}} = \frac{\pi K}{4} \log\left(\frac{R}{r_{\text{core}}}\right). \tag{8.10}$$

Here, R is the system radius, r_{core} is the disclination core radius, and we have used the winding number $q = \pm 1/2$.

If those assumptions are not valid, then Eq. (8.10) provides a reasonable first approximation. A more precise free energy calculation then becomes much more complex. In particular, unequal Frank constants will change the geometric structure of the disclination, so that the director configuration will have more of the lower-free-energy elastic modes and less of the higher-free-energy elastic modes. Also, varying the $\hat{\boldsymbol{t}}$ or $\hat{\boldsymbol{\Omega}}$ vectors will presumably cost extra free energy. We do not know of any analytic theory for those effects, but they can be studied through numerical simulations.

When a disclination line moves, it experiences a drag force per length. For a simple model of the drag, we can follow the same argument as in Sect. 4.1.1. Here, we assume that the Rayleigh dissipation function is

[7] In Sect. 7.3, we discussed the elastic free energy of a 2D nematic liquid crystal, which can have two distinct Frank elastic constants. By comparison, a 3D nematic liquid crystal can have four distinct Frank elastic constants: K_{11}, K_{22}, K_{33}, and K_{24}. The relatively simple free energy of Eq. (8.9) applies if $K_{11} = K_{22} = K_{33} = 2K_{24} \equiv K$. See J. V. Selinger, "Interpretation of Saddle-Splay and the Oseen-Frank Free Energy in Liquid Crystals," *Liq. Cryst. Rev.* **6**, 129 (2018).

$$D = \frac{1}{2}\gamma \int d\boldsymbol{r} |\dot{\boldsymbol{n}}|^2, \tag{8.11}$$

with only a single rotational viscosity γ. Equation (4.10) then gives the drag force per length

$$\boldsymbol{f}_{\text{drag}} = -\left[\frac{\pi\gamma}{4}\log\left(\frac{R}{r_{\text{core}}}\right)\right]\boldsymbol{u}, \tag{8.12}$$

where $\boldsymbol{u}$ is the velocity of the moving disclination line, and we have again used $q = \pm 1/2$. The calculation becomes much more complex if there are several viscosity coefficients, which couple director rotation and fluid flow. In general, those problems must be studied numerically.

If a system has more than one disclination line, then the disclination lines interact with each other through elastic distortions in the director field. The interaction may be either repulsive or attractive. A recent paper has investigated this interaction, assuming a single Frank elastic constant.[8] It argues that the interaction force can be expressed as

$$\boldsymbol{f}_{12} = \frac{\pi K}{2}(\hat{\boldsymbol{\Omega}}_1 \cdot \hat{\boldsymbol{\Omega}}_2)(\hat{\boldsymbol{t}}_1 \cdot \hat{\boldsymbol{t}}_2)\frac{\boldsymbol{r}_{12} L}{|\boldsymbol{r}_{12}|^2}, \quad \text{if disclinations are parallel or antiparallel,}$$

$$\boldsymbol{f}_{12} = \frac{\pi^2 K}{2}\frac{(\hat{\boldsymbol{\Omega}}_1 \cdot \hat{\boldsymbol{\Omega}}_2)(\hat{\boldsymbol{t}}_1 \cdot \hat{\boldsymbol{t}}_2)}{[1-(\hat{\boldsymbol{t}}_1 \cdot \hat{\boldsymbol{t}}_2)^2]^{1/2}}\frac{\boldsymbol{r}_{12}}{|\boldsymbol{r}_{12}|}, \quad \text{otherwise,} \tag{8.13}$$

where L is the length along the disclinations. Hence, the interaction is repulsive if $(\hat{\boldsymbol{\Omega}}_1 \cdot \hat{\boldsymbol{\Omega}}_2)(\hat{\boldsymbol{t}}_1 \cdot \hat{\boldsymbol{t}}_2) > 0$, and attractive if $(\hat{\boldsymbol{\Omega}}_1 \cdot \hat{\boldsymbol{\Omega}}_2)(\hat{\boldsymbol{t}}_1 \cdot \hat{\boldsymbol{t}}_2) < 0$. This criterion for repulsion or attraction is consistent with the Coulomb-like interaction presented in Sect. 3.2. In a 2D system, we can define $\hat{\boldsymbol{t}} = \hat{\boldsymbol{z}}$ for all disclinations, $\hat{\boldsymbol{\Omega}} = \hat{\boldsymbol{z}}$ for $+1/2$ disclinations, and $\hat{\boldsymbol{\Omega}} = -\hat{\boldsymbol{z}}$ for $-1/2$ disclinations. Equation (8.13) then implies that the interaction is repulsive for charges of the same sign, and attractive for charges of opposite signs, as in Coulomb's law. It further shows how the interaction varies for any vectors $\hat{\boldsymbol{t}}$ and $\hat{\boldsymbol{\Omega}}$. As usual, the generalization to unequal Frank constants is much more complex, and must be done numerically.

8.4.4 Variable Magnitude of Order

We have previously discussed how the magnitude of orientational order varies inside the core of a topological defect. In particular, Sect. 5.1 presented the theory for 2D polar order, and Sect. 7.1.3 presented the theory for 2D nematic order. In both of those cases, the orientational order goes to zero at the center of a defect.

---

[8] C. Long, X. Tang, R. L. B. Selinger, and J. V. Selinger, "Geometry and Mechanics of Disclination Lines in 3D Nematic Liquid Crystals," *Soft Matter* **17**, 2265 (2021).

Here, we introduce an analogous theory for 3D nematic order.[9] The concept is similar, but the results are somewhat more complex than the two previous cases.

In general, 3D nematic order is represented by a tensor field $Q_{ij}(\boldsymbol{r})$, which is symmetric and traceless.[10] At any position $\boldsymbol{r}$, this tensor has three eigenvectors, which describe the principal axes of orientational order at that position. It also has three corresponding eigenvalues, which describe the local magnitude of order along each of those axes. Because the tensor is traceless, the three eigenvalues add up to zero.

Away from any defects, the order is *uniaxial*, meaning that the director $\hat{\boldsymbol{n}}$ is the only special orientation in space, and the two directions perpendicular to $\hat{\boldsymbol{n}}$ are equivalent to each other. In that case, three eigenvalues can be written as S (corresponding to eigenvector $\hat{\boldsymbol{n}}$), along with $-\frac{1}{2}S$ and $-\frac{1}{2}S$ (corresponding to the two directions perpendicular to $\hat{\boldsymbol{n}}$). Here, S is called the scalar order parameter. The tensor Q_{ij} can then be expressed in terms of S and $\hat{\boldsymbol{n}}$ as

$$Q_{ij} = S\left(\frac{3}{2}n_i n_j - \frac{1}{2}\delta_{ij}\right). \tag{8.14}$$

Inside the core of a defect, all three eigenvalues of Q_{ij} may vary, while keeping their sum equal to zero. In particular, this tensor does not need to remain uniaxial. It may become locally *biaxial*, meaning that all three principal axes are different from each other. In this small region, Eq. (8.14) is not valid. Instead, we must develop a more general expression for Q_{ij}.

To develop a more general expression, suppose we have a disclination running along the z-axis, and use cylindrical coordinates with $x = \rho\cos\phi$, $y = \rho\sin\phi$. As we move in a loop around the z-axis, the director $\hat{\boldsymbol{n}}$ rotates through an angle of π about the rotation axis $\hat{\boldsymbol{\Omega}}$. Hence, it can be written as

$$\hat{\boldsymbol{n}} = \hat{\boldsymbol{m}}\cos\left(\frac{\phi}{2}\right) + \hat{\boldsymbol{m}}'\sin\left(\frac{\phi}{2}\right), \tag{8.15}$$

where $\hat{\boldsymbol{m}}$ and $\hat{\boldsymbol{m}}'$ are two orthonormal vectors in the plane perpendicular to $\hat{\boldsymbol{\Omega}}$. The three principal axes of Q_{ij} are $\hat{\boldsymbol{n}}$, $\hat{\boldsymbol{\Omega}}$, and

$$\hat{\boldsymbol{n}}_\perp = -\hat{\boldsymbol{m}}\sin\left(\frac{\phi}{2}\right) + \hat{\boldsymbol{m}}'\cos\left(\frac{\phi}{2}\right). \tag{8.16}$$

A general expression for the nematic order tensor then becomes

[9] The theory in this section was developed by N. Schopohl and T. J. Sluckin, "Defect Core Structure in Nematic Liquid Crystals," *Phys. Rev. Lett.* **59**, 2582 (1987).

[10] For further background on the tensor order parameter for a nematic phase, see our previous book J. V. Selinger, *Introduction to the Theory of Soft Matter: From Ideal Gases to Liquid Crystals* (Springer, 2016), Section 10.2.

$$Q_{ij} = Sn_i n_j + S_\perp n_{\perp i} n_{\perp j} - (S + S_\perp)\Omega_i \Omega_j, \tag{8.17}$$

with eigenvalues S, $S_\perp$, and $-(S+S_\perp)$, which may all depend on position. Far from the defect, the liquid crystal becomes uniaxial, and we must have $S_\perp = -\frac{1}{2}S$.

In Landau-de Gennes theory, the free energy of a nematic liquid crystal is written as a power series in the order tensor Q_{ij}. One conventional expression is

$$F = \int d\boldsymbol{r}\left[\frac{A}{2}\left(1 - \frac{\gamma}{3}\right) Q_{ij} Q_{ij} - \frac{A\gamma}{3} Q_{ij} Q_{jk} Q_{ki} + \frac{A\gamma}{4}(Q_{ij} Q_{ij})^2 + \frac{L}{2}(\partial_k Q_{ij})(\partial_k Q_{ij})\right]. \tag{8.18}$$

We put the nematic order tensor of Eq. (8.17) into the free energy of Eq. (8.18), and obtain the free energy in terms of S and $S_\perp$. We derive the Euler-Lagrange equations, and solve them numerically, to obtain $S(\rho)$ and $S_\perp(\rho)$ as functions of the distance ρ from the center of the disclination.

As an example, Fig. 8.6 presents numerical results for all three eigenvalues of Q_{ij} as functions of position around a disclination, with the parameter $\gamma = 4.8$. Far from the disclination, the system is uniaxial with $S = 0.5$. Hence, in the bulk, the three eigenvalues are 0.5 (for $\hat{\boldsymbol{n}}$), -0.25 (for $\hat{\boldsymbol{n}}_\perp$), and -0.25 (for $\hat{\boldsymbol{\Omega}}$). As we move in toward the center of the disclination, within a core radius of order $\rho_{\text{core}} = \sqrt{L/A}$, the eigenvalue for $\hat{\boldsymbol{n}}$ becomes lower, and the eigenvalue for $\hat{\boldsymbol{n}}_\perp$ becomes higher, while the eigenvalue for $\hat{\boldsymbol{\Omega}}$ remains approximately constant. In this core region, the nematic order is biaxial, with three distinct eigenvalues. Right at the center, the eigenvalues for $\hat{\boldsymbol{n}}$ and $\hat{\boldsymbol{n}}_\perp$ cross each other. Because those two eigenvalues are equal, the nematic order at the center is not biaxial, but rather *negative uniaxial nematic*—meaning that the molecules have a uniaxial distribution, predominantly *away from* $\hat{\boldsymbol{\Omega}}$.

Hence, the central axis of the disclination is not a place with zero orientational order. Rather, it is a place where two of the eigenvalues cross, so that the principal axes $\hat{\boldsymbol{n}}$ and $\hat{\boldsymbol{n}}_\perp$ are exchanged.

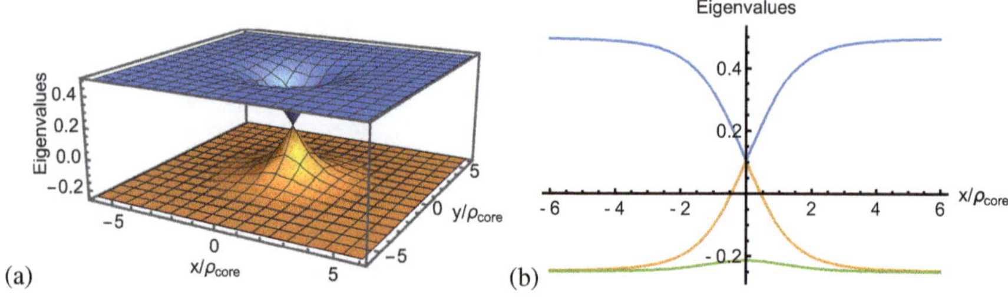

Fig. 8.6 Numerical solution for the eigenvalues of the nematic order tensor, showing how the highest two eigenvalues cross at the center of a disclination. (**a**) 3D plot of the eigenvalues as functions of x and y. Only the highest two eigenvalues are visible; the lowest eigenvalue is hidden. (**b**) 2D plot of the eigenvalues as functions of x, for $y = 0$. Here, all three eigenvalues are visible

8.5 What Happens to Topological Charge and Charge Density?

In the previous chapters, for 2D polar order and 2D nematic order, we saw that winding number serves as a topological charge, analogous to electric charge. For 2D polar order, it must be an integer. For 2D nematic order, it must be an integer or half-integer. In either case, the topological charges combine in a very simple, straightforward way—just like addition of ordinary integers or half-integers. In a mathematical sense, this addition rule arises from the statement that $\pi_1(S^1) = \mathbb{Z}$, the set of integers, in Eq. (8.3).

Now, for 3D nematic order, the concept of topological charge becomes more complicated. For 3D nematic order, the loop construction shows only two types of director configurations: defect-free and half-charge disclination. Hence, the only possible topological charges are 0 and $1/2$. These charges do not combine with the rules of ordinary addition. Rather, they combine as

$$0 + 0 = 0, \qquad 0 + \frac{1}{2} = \frac{1}{2}, \qquad \frac{1}{2} + \frac{1}{2} = 0. \tag{8.19}$$

Those rules can be described as addition of half-integers, modulo 1. They are similar to the concept of parity (in physics or computer science). They arise from the statement that $\pi_1(S^2/\mathbb{Z}_2) = \pi_1(\mathbb{R}P^2) = \mathbb{Z}_2$, a set with two elements, in Eq. (8.5).

For an explicit illustration of these addition rules, see Fig. 8.7. This figure shows a simulation of a 3D nematic phase inside a cylinder, with boundary conditions that require the director to be perpendicular to the curved wall. This geometry is commonly used in experiments on liquid crystals. In response to the boundary conditions, the director field has formed a disclination loop, shown in red. The structure of this loop is the same as the example in Fig. 8.5a, but rotated by 90°.

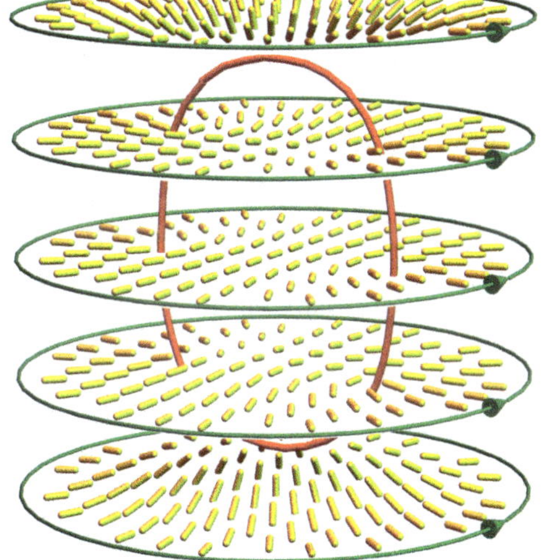

Fig. 8.7 Simulation of a 3D nematic phase inside a cylinder, with boundary conditions that anchor the director perpendicular to the curved wall. The director field forms a disclination loop, shown in red. In the middle three cross sections, we see two disclinations with $+1/2$ wedge structure. In the top and bottom cross sections, the director field escapes into the third dimension, and has no defects. The simulation code was kindly provided by Cheng Long

Here, we want to emphasize the structure in successive circular cross sections through the cylinder:

- In the top cross section, the green loop encloses a winding number of $+1$, and there are no defects. Rather, the director field with this winding number escapes into the third dimension, as discussed in Sect. 8.1. This configuration is called *escaped radial*, with splay pointing up.
- In the middle cross sections, the green loop encloses a winding number of $+1$, and there are two disclinations. Each of these disclinations has a $+1/2$ wedge structure. This configuration is called *planar polar*.
- In the bottom cross section, the green loop still encloses a winding number of $+1$, and there are no defects. The configuration is escaped radial, but now with splay pointing down.

As a function of height z, the system begins with no defects, then two $+1/2$ wedge disclinations form and move apart. Farther down, these $+1/2$ wedge disclinations move together and annihilate each other, leaving behind a defect-free configuration.

This structure can be compared with Fig. 7.4b. In that case, the system begins with no defects, then a pair of $+1/2$ and $-1/2$ wedge disclinations form and move apart. Farther down, the $+1/2$ and $-1/2$ wedge disclinations move together and annihilate each other, leaving a defect-free configuration.

If the director field is free to rotate in 3D, then both of these creation and annihilation processes can occur. By contrast, if the director field is confined to 2D, then the defects must be created and annihilated in pairs of positive and negative.

Along with topological charge, we can also mention charge density. In Sect. 6.2.1, we defined a concept of topological charge density for defects in 2D polar order. The same concept also works for 2D nematic order. However, it becomes more complicated for 3D nematic order. Indeed, because the ordinary addition of topological charges does not work for 3D nematic order, and integration of charge density is a limiting case of addition, we cannot really expect it to work in the ordinary way.

A recent paper has suggested a way of defining defect charge density for 3D nematic order.[11] Essentially, this paper generalizes Eq. (6.19) into the tensor

$$D_{ab}(\boldsymbol{r}) = \frac{1}{2\pi}\epsilon_{alk}\partial_l(\epsilon_{bij}n_i\partial_k n_j). \tag{8.20}$$

In any region of uniaxial nematic order with constant magnitude, that definition is equivalent to

[11] C. D. Schimming and J. Viñals, "Singularity Identification for the Characterization of Topology, Geometry, and Motion of Nematic Disclination Lines," *Soft Matter* **18**, 2234 (2022). In Eq. (8.21), we have modified their definition by a factor of $(2/3)^2$ to be consistent with our definition of the nematic order tensor, and by a factor of $1/2$ because of our convention that the topological charge of a disclination should be $1/2$.

$$D_{ab}(\boldsymbol{r}) = \frac{1}{2\pi}\left(\frac{2}{3S_0}\right)^2 \epsilon_{alk}\partial_l(\epsilon_{bij}Q_{ic}\partial_k Q_{jc}) = \frac{2}{9\pi S_0^2}\epsilon_{alk}\epsilon_{bij}(\partial_l Q_{ic})(\partial_k Q_{jc}). \tag{8.21}$$

Hence, the paper suggests to use Eq. (8.21) in terms of the nematic order tensor everywhere.

This definition of a tensor charge density has several desirable features: *If a disclination is present,* then Eq. (8.20) has a δ-function spike at the disclination. Likewise, Eq. (8.21) has a nonsingular peak at the disclination, broadened over the core radius (analogous to the broadening shown in Fig. 6.29). Furthermore, the tensor D_{ab} has the tensor structure of $\Omega_a t_b$, so that it can be used to determine the rotation vector $\hat{\boldsymbol{\Omega}}$ and tangent vector $\hat{\boldsymbol{t}}$.

However, as the same authors noted in a subsequent paper,[12] this definition also has a serious problem: It has *false positives*, meaning that it is nonzero in regions where the director configuration escapes into the third dimension, where no disclinations are present. For example, in the top and bottom cross sections of Fig. 8.7, the tensor D_{ab} is nonzero everywhere (using either Eq. (8.20) or Eq. (8.21)).

Based on these considerations, we would say that the tensor D_{ab} may be quite useful for analyzing simulations, but it cannot be strictly regarded as a topological charge density. Indeed, we are not sure whether 3D nematic order can have any topological charge density, which can capture the difference between a disclination and escape into the third dimension. This distinction is a global concept, which might or might not be describable locally.

[12] C. D. Schimming and J. Viñals, "A Tensor Density Measure of Topological Charge in Three-Dimensional Nematic Phases," *Proc. R. Soc. A* **480**, 20230564 (2024).

Defects in Crystals

9

In the previous chapters, we have considered systems with orientational order, such as liquid crystals and magnets. We have investigated the topological defects and solitons that can occur in such systems. In this chapter, we will extend the discussion to defects in crystals. We will see that crystals can have two types of topological defects, called *dislocations* and *disclinations*, which are both closely related to the topological defects that we have discussed earlier.

Dislocations are extremely common in crystals, and they are extremely important for technology. As an example, the strength of materials is generally determined by the microstructure of dislocations. For that reason, dislocations are widely studied in metallurgy and other areas of materials science, and there is a vast scientific literature about dislocations. Disclinations are much less common, and hence less studied in the literature, but they are still part of the overall research field on defects in crystalline materials.

In this chapter, we will not be able cover this huge scientific field in any detail. Rather, we will just describe some of the fundamental features of dislocations and disclinations, and we will emphasize how they compare with the defects that we have studied previously. If any readers would like further detail, they should consult a full textbook about crystalline defects.[1]

9.1 Translational and Orientational Order in Crystals

In general, the concept of *order* refers to *spontaneous symmetry breaking*. As a system cools, it goes through a phase transition from a higher-temperature phase, which has *more symmetry*, to a lower-temperature phase, which has *less symmetry*. In this transition, the system could go into many possible states, and it randomly

[1] For one excellent textbook, see P. M. Anderson, J. P. Hirth, and J. Lothe, *Theory of Dislocations*, third edition (Cambridge University Press, 2017).

J. V. Selinger, *Introduction to Topological Defects and Solitons*, Lecture Notes in Physics 1032, https://doi.org/10.1007/978-3-031-70200-6_9

chooses one of the possibilities. For example, a ferromagnet could align in any direction, and it randomly chooses one direction $\hat{\boldsymbol{n}}$. Likewise, a nematic liquid crystal could align along any axis, and it randomly chooses one axis $\pm\hat{\boldsymbol{n}}$.

In crystals, there are two types of order:

- *Translational* (also called *positional*)
- *Orientational* (also called *bond-orientational*)

It is important to distinguish between these two types of order, because each type of order is associated with a different type of topological defect.

Translational order refers to order in the positions of the particles. It means that a system of particles has broken translational symmetry, and randomly selected a specific set of positions on a regular lattice. For example, suppose that some material tends to form a 2D square lattice, as shown in Fig. 9.1. When the material cools from the liquid phase into the crystal phase, it might form a crystal with the particles on the black dots in Fig. 9.1a. Alternatively, it might form a crystal with the particles on the orange dots, which are uniformly displaced from the black dots. Indeed, it might form a crystal with the particles uniformly displaced from the black dots by any distance in any direction. All of these possibilities have exactly the same free energy. Translational order is the choice of one of these possibilities, and not the others.

Orientational order means that a system of particles has broken rotational symmetry, and randomly selected a specific set of directions. To see the orientational

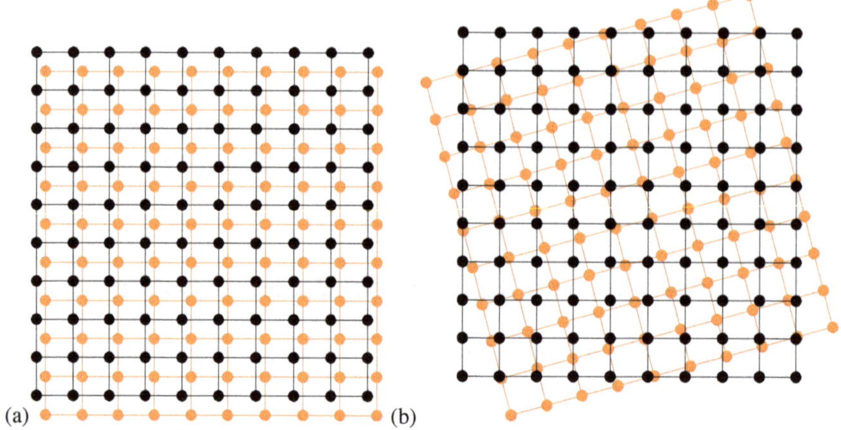

Fig. 9.1 (**a**) Translational order in a 2D square lattice. When the material crystallizes, it randomly chooses the black set of points, or the orange set of points, or any other set of points that are uniformly displaced. (**b**) Orientational order of the bonds in a 2D square lattice. When the material crystallizes, it randomly chooses the black bond orientations, or the orange bond orientations, or any other bond orientations that are uniformly rotated

order, we must consider the bonds between particles in a crystal.[2] Again, suppose that some material tends to form a 2D square lattice. When it cools from the liquid phase into the crystal phase, it might form a crystal with the nearest-neighbor bonds along the x and y-axes, as shown in black in Fig. 9.1b. Alternatively, it might form a crystal with the bonds at 15° from the x and y-axes, as shown in orange. It might form a crystal with the bonds rotated by any angle. All of those possibilities have exactly the same free energy. Orientational order is the choice of one of those possibilities, and not the others.

For the black and orange lattices in Fig. 9.1a, the positions of the particles are different, but the orientations of the bonds are the same. This example shows that we can change translational order without changing orientational order. By contrast, for the black and orange lattices in Fig. 9.1b, the orientations of the bonds are different, and the positions of the particles are completely different. That example shows that we cannot change orientational order without also changing translational order.

9.2 Dislocations

A dislocation is a defect that disrupts the translational order of a crystal, without changing the orientational order. We discuss the structure of dislocations first in 2D, and then in 3D.

9.2.1 Dislocation Structure in 2D

Let us begin with a 2D square lattice. Figure 9.2a shows a perfect square lattice, and Fig. 9.2b shows a square lattice with a dislocation, indicated by the red dot. In Fig. 9.2b, it looks as if someone has removed a half-row of particles, running from the dislocation along the negative x-axis, and then allowed the particles to shift to fill in the gap. (Alternatively, it looks as if someone has inserted a half-row of particles, running from the dislocation along the positive x-axis, and then allowed the particles to shift to reduce overcrowding.) Near the dislocation, the lattice is highly distorted. As we move away from the dislocation, the distortions gradually decrease. Hence, the structure is a *point defect in 2D*. There is no defect along the half-row of particles that has been removed (or inserted).

To characterize the dislocation, we can use a loop construction, which is very similar to the loop construction for topological defects in previous chapters. First, consider the perfect crystal in Fig. 9.2a. Suppose we follow the green loop: begin in the lower right, move up by 5 lattice units, left by 5 units, down by 5 units, and right by 5 units. At the end, we are back in exactly the same place where we began. Now,

[2] In this context, the word *bond* just means the line between neighboring particles. This nearest-neighbor bond might or might not be a covalent bond. In the current discussion, we are only concerned with the geometry, not with the chemical nature of the bond.

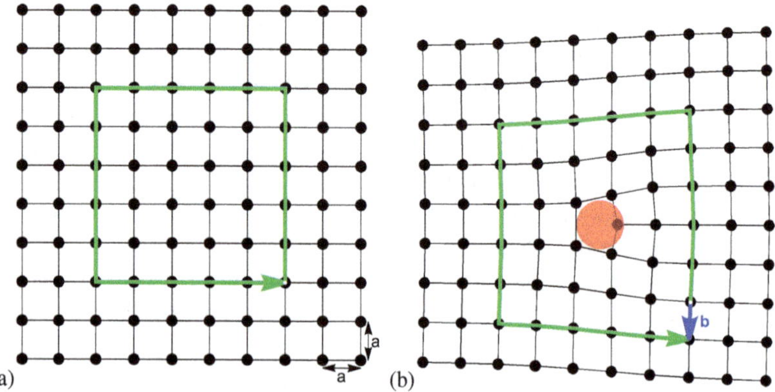

Fig. 9.2 (**a**) Perfect 2D square lattice. The green Burgers circuit begins and ends at exactly the same point. (**b**) Square lattice with a dislocation, shown by the red disk. The green Burgers circuit does not end at the same point where it begins. The difference between these points is the Burgers vector, shown in blue

consider the crystal with a dislocation in Fig. 9.2b. We follow the same loop: up by 5 units, left by 5 units, down by 5 units, and right by 5 units. At the end, we are *not* back where we began. Rather, we are in a different position, which is shifted from the starting point by one unit downward.

The loop is called a *Burgers circuit*, and the shift from beginning to end is called the *Burgers vector* $\boldsymbol{b}$. In this example, $\boldsymbol{b} = -a\hat{\mathbf{y}}$, where a is the lattice constant of the square lattice. It is indicated by the blue vector in the figure.

This loop construction gives the same Burgers vector $\boldsymbol{b}$ for any counter-clockwise loop that encloses the dislocation, and it gives zero for any loop that does not enclose the dislocation. It does not matter whether the loop is large or small, or whether it is square or rectangular or another shape. We only need to draw a loop that *should* close in a perfect crystal, and determine whether it *actually* closes in the crystal with a dislocation. Readers may wish to practice drawing a few loops to check this statement.

Note that the Burgers vector goes from one lattice site to another lattice site. Hence, it must be a vector in the lattice. In our current example, we have a 2D square lattice with lattice constant a, and the Burgers vector must be $\boldsymbol{b} = ia\hat{\mathbf{x}} + ja\hat{\mathbf{y}}$, where i and j are integers. Only a vector of that form can go from one lattice site to another lattice site.

The concept of a dislocation is not limited to square lattices. For example, consider a 2D hexagonal lattice. In the perfect lattice, shown in Fig. 9.3a, we can follow the green loop, moving 3 units along each side of a hexagon. At the end of the loop, we are exactly back at the starting point. By contrast, in the lattice with a dislocation, shown in Fig. 9.3b, we can follow the same green loop, but we do not return to the starting point. Rather, the end point is shifted from the starting point by one unit downward. Hence, the Burgers vector is again $\boldsymbol{b} = -a\hat{\mathbf{y}}$. We would get the

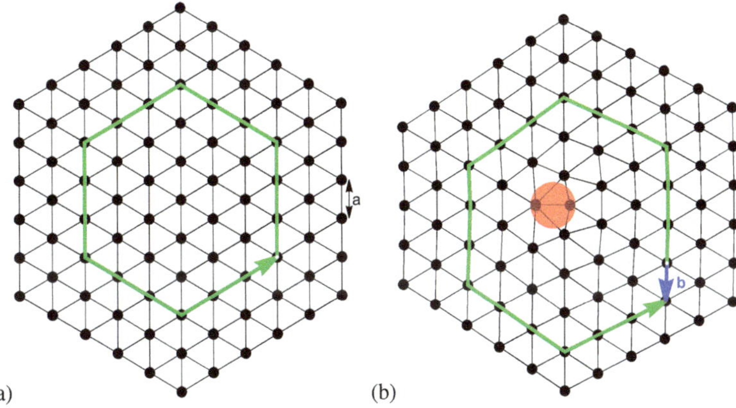

Fig. 9.3 (**a**) Perfect 2D hexagonal lattice. The green Burgers circuit begins and ends at the same point. (**b**) Hexagonal lattice with a dislocation, shown by the red disk. The green Burgers circuit does not end at the same point where it begins. The difference between these points is the Burgers vector, shown in blue

same Burgers vector for any counter-clockwise loop around this dislocation—large or small, hexagon or parallelogram or other shape.

In this 2D hexagonal lattice, the Burgers vector must be compatible with the lattice type. It can be written as $\boldsymbol{b} = i\boldsymbol{v}_1 + j\boldsymbol{v}_2$, where $\boldsymbol{v}_1$ and $\boldsymbol{v}_2$ are lattice vectors, $\boldsymbol{v}_{1,2} = a(\hat{\boldsymbol{x}} \cos 30^\circ \pm \hat{\boldsymbol{y}} \sin 30^\circ)$, and i and j are integers.

In Figs. 9.2b and 9.3b, we have drawn the dislocations as broad red disks, rather than as mathematical points. The reason for this symbol is that we cannot resolve the position of a dislocation on a length scale finer than the lattice constant a. The broad red disk represents a core region, analogous to the defect cores discussed in Sects. 5.1, 7.1.3, and 8.4.4. Within the core, particles do not have the normal number of nearest neighbors (four for a square lattice, six for a hexagonal lattice). Hence, if we tried to construct a Burgers circuit through the core, there would be some ambiguity in how to draw the line. Outside the core, particles have the normal number of nearest neighbors, so there is no ambiguity in how to draw a Burgers circuit. We will return to this topic of dislocation core structure in Sect. 9.4.1.

9.2.2 Dislocation Structure in 3D

The concept of a dislocation can be extended from a *defect point* in 2D to a *defect line* in 3D. As an example, consider a 3D simple cubic lattice. Figure 9.4a shows a perfect version of this lattice. Here, the green Burgers circuit closes perfectly, indicating that there is no Burgers vector inside the loop. By contrast, Fig. 9.4b shows a simple cubic lattice with a dislocation. In this case, the structure in Fig. 9.2b is just extended uniformly upward in z. The dislocation point in 2D becomes a

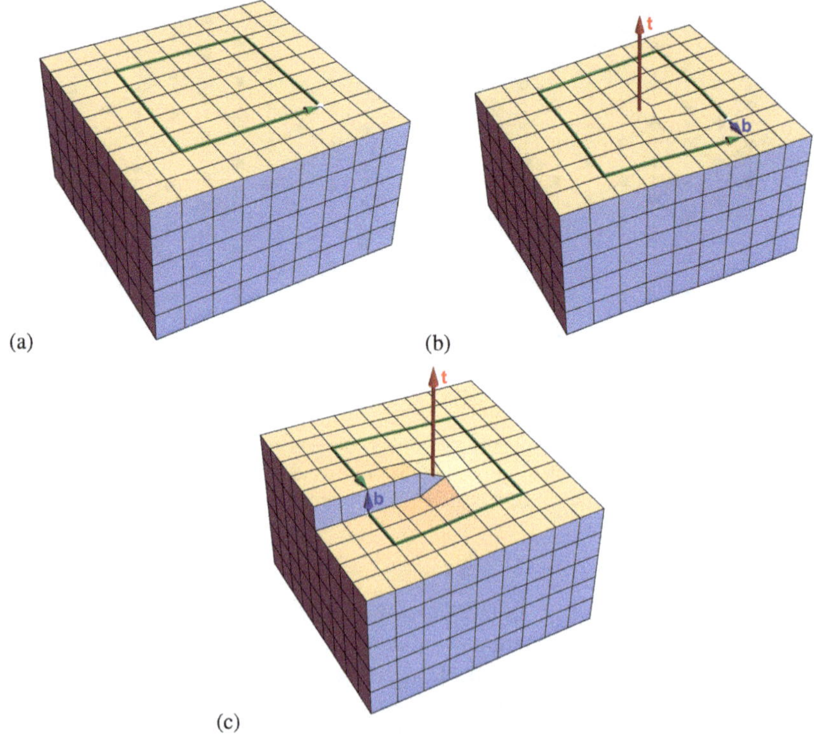

Fig. 9.4 (**a**) Perfect 3D simple cubic lattice, with the green Burgers circuit beginning and ending at the same point. (**b**) Simple cubic lattice with an edge dislocation. The blue Burgers vector is perpendicular to the red tangent vector. (**c**) Simple cubic lattice with a screw dislocation. The blue Burgers vector is parallel to the red tangent vector

vertical red line in 3D, with the tangent vector $\hat{\boldsymbol{t}} = \hat{z}$ pointing upward.[3] The green Burgers circuit runs around the dislocation line, in the direction determined by $\hat{\boldsymbol{t}}$ through the right-hand rule. The end point of this circuit is shifted from the starting point by one lattice unit in the negative y direction. Hence, the Burgers vector is $\hat{\boldsymbol{b}} = -a\hat{y}$, shown in blue. Any other Burgers circuit around the dislocation line would give the same Burgers vector.

In the example of Fig. 9.2b, the vectors $\hat{\boldsymbol{t}}$ and $\hat{\boldsymbol{b}}$ are *perpendicular* to each other. The resulting structure is called an *edge dislocation*.

As an alternative possibility, in a 3D crystal, the vectors $\hat{\boldsymbol{t}}$ and $\hat{\boldsymbol{b}}$ might be *parallel* or *antiparallel* to each other. In that case, the structure is called a *screw dislocation*. An example is shown in Fig. 9.2c. This structure looks like a parking garage. If we drive around a Burgers circuit (in the direction determined by $\hat{\boldsymbol{t}}$ through the

[3] In the literature about dislocations, the tangent vector is often labeled as $\hat{\boldsymbol{\xi}}$. However, in this chapter, we use the symbol $\hat{\boldsymbol{t}}$ to emphasize the analogy with disclination lines in Chap. 8.

right-hand rule), our end point is shifted to one lattice unit above the starting point. Hence, the Burgers vector is $\hat{\boldsymbol{b}} = a\hat{\mathbf{z}}$, parallel to $\hat{\boldsymbol{t}}$. The structure is helical, with a right-handed twist. If $\hat{\boldsymbol{t}}$ and $\hat{\boldsymbol{b}}$ were antiparallel, we would have the mirror-image helix, with a left-handed twist.

In general, the Burgers vector $\hat{\boldsymbol{b}}$ might have components perpendicular and parallel to the tangent vector $\hat{\boldsymbol{t}}$. In that case, the dislocation would have both edge and screw components.

As in Sect. 8.4.2, the definition of $\hat{\boldsymbol{t}}$ requires one arbitrary choice: We must decide which way is forward or backward along the dislocation line. If we switch the choice of forward and backward, then $\hat{\boldsymbol{t}}$ would change to $-\hat{\boldsymbol{t}}$. Because $\hat{\boldsymbol{t}}$ determines which way to move around the Burgers circuit, the beginning and end of this circuit would switch, and hence $\hat{\boldsymbol{b}}$ would change to $-\hat{\boldsymbol{b}}$. Our description of the dislocation must be invariant under this transformation. As an aside, we should also note that the definition of $\hat{\boldsymbol{b}}$ involves using the right-hand rule once, so $\hat{\boldsymbol{b}}$ is a pseudovector.

In some ways, the Burgers vector $\boldsymbol{b}$ for a dislocation line is analogous to $\pi\hat{\boldsymbol{\Omega}}$, the rotation vector multiplied by π, for a nematic disclination line. If we move in a loop around a nematic disclination line, the director rotates by the magnitude π in the direction described by $\hat{\boldsymbol{\Omega}}$. Likewise, if we move in a loop around a dislocation line, the crystalline lattice is displaced by $|\boldsymbol{b}|$ in the direction described by $\boldsymbol{b}/|\boldsymbol{b}|$. Furthermore, the geometric structure of a nematic disclination is characterized by the angle between $\hat{\boldsymbol{t}}$ and $\hat{\boldsymbol{\Omega}}$, just as the geometric structure of a dislocation is characterized by the angle between $\hat{\boldsymbol{t}}$ and $\boldsymbol{b}$.

However, there is one important difference between $\hat{\boldsymbol{\Omega}}$ and $\boldsymbol{b}$. For a nematic disclination, $\hat{\boldsymbol{\Omega}}$ can be any unit vector. Hence, it has a continuous set of possibilities. By contrast, $\boldsymbol{b}$ must go from one lattice site to another lattice site, so it must be a vector in the lattice. For example, in the simple cubic lattice, it must have the form $\boldsymbol{b} = ia\hat{\mathbf{x}} + ja\hat{\mathbf{y}} + ka\hat{\mathbf{z}}$. This is only a discrete set of possibilities.

A dislocation line does not need to be straight. Like a nematic disclination line, a dislocation line may curve. It cannot begin or end anywhere inside a crystal. Instead, it might go all the to the boundary of the crystal, or it might curve into a closed loop. If the dislocation line curves, then the tangent vector $\hat{\boldsymbol{t}}$ must vary continuously along the line. However, the Burgers vector $\boldsymbol{b}$ *cannot* vary continuously along the line, because it has only a discrete set of possibilities. In any topological defect, the properties cannot jump discontinuously. Hence, $\boldsymbol{b}$ must remain fixed; it does not change either magnitude or orientation along a dislocation line.

Based on that argument, the vectors $\hat{\boldsymbol{t}}$ and $\boldsymbol{b}$ *cannot* rotate together, as $\hat{\boldsymbol{t}}$ and $\hat{\boldsymbol{\Omega}}$ rotate together in Fig. 8.5a–c. Rather, $\hat{\boldsymbol{t}}$ can change while $\boldsymbol{b}$ remains fixed. In that case, the angle between $\hat{\boldsymbol{t}}$ and $\boldsymbol{b}$ will change, and hence the edge or screw structure of the dislocation will change.

9.2.3 Burgers Vector as Topological Charge

Let us develop a mathematical expression for the loop construction that defines the Burgers vector $\boldsymbol{b}$ of a dislocation. Outside the core region, we can regard the crystal structure as a deformed version of a perfect crystal. The perfect crystal can be described by coordinates $\boldsymbol{r} = (x, y, z)$, and the deformed crystal by coordinates $\boldsymbol{r}' = (x', y', z')$. The deformed position of any particle is shifted from the undeformed position by

$$\boldsymbol{r}'(\boldsymbol{r}) = \boldsymbol{r} + \boldsymbol{u}(\boldsymbol{r}), \tag{9.1}$$

where $\boldsymbol{u}(\boldsymbol{r})$ is the *displacement field*. The loop construction can then be written as an integral in the deformed coordinates as

$$\boldsymbol{b} = \oint d\boldsymbol{r}'. \tag{9.2a}$$

This expression can be transformed into an integral in the undeformed coordinates as

$$\boldsymbol{b} = \oint d\boldsymbol{r} + \oint d\boldsymbol{u}. \tag{9.2b}$$

The first term $\oint d\boldsymbol{r}$ vanishes because the Burgers circuit must close in a perfect crystal. Hence, we are left with

$$\boldsymbol{b} = \oint d\boldsymbol{u}. \tag{9.3a}$$

In terms of vector components, that equation is

$$b_i = \oint du_i = \oint (\nabla u_i) \cdot d\boldsymbol{l} = \oint (\partial_j u_i) dl_j. \tag{9.3b}$$

Equation (9.3) for the Burgers vector is analogous to Eq. (1.8) for topological charge in the xy model. Hence, the Burgers vector is a *vector topological charge* associated with a dislocation.

In previous chapters, we have seen that topological charges combine in different ways, depending on the type of order. For 2D polar order and 2D nematic order, topological charges combine in a simple way, as ordinary addition of integers or half-integers. By contrast, for 3D nematic order, topological charges combine in a more subtle way, as addition of half-integers, modulo 1.

Now, for translational order in crystals, Burgers vectors combine following the ordinary rules for addition of vectors. If a Burgers circuit goes around two dislocations with Burgers vectors $\boldsymbol{b}_1$ and $\boldsymbol{b}_2$, it shows a total Burgers vector of $\boldsymbol{b}_1 + \boldsymbol{b}_2$. As an example, Fig. 9.5a shows a 2D square lattice with two edge dislocations. The small green Burgers circuits show that each dislocation has a Burgers vector of $\boldsymbol{b}_1 = \boldsymbol{b}_2 = -a\hat{\boldsymbol{y}}$. The large Burgers circuit shows that the total Burgers vector is $\boldsymbol{b}_1 + \boldsymbol{b}_2 = -2a\hat{\boldsymbol{y}}$. For comparison, Fig. 9.5b shows a 2D square lattice with edge dislocations of Burgers vector $\boldsymbol{b}_1 = a\hat{\boldsymbol{x}}$ and $\boldsymbol{b}_2 = -a\hat{\boldsymbol{y}}$. The large

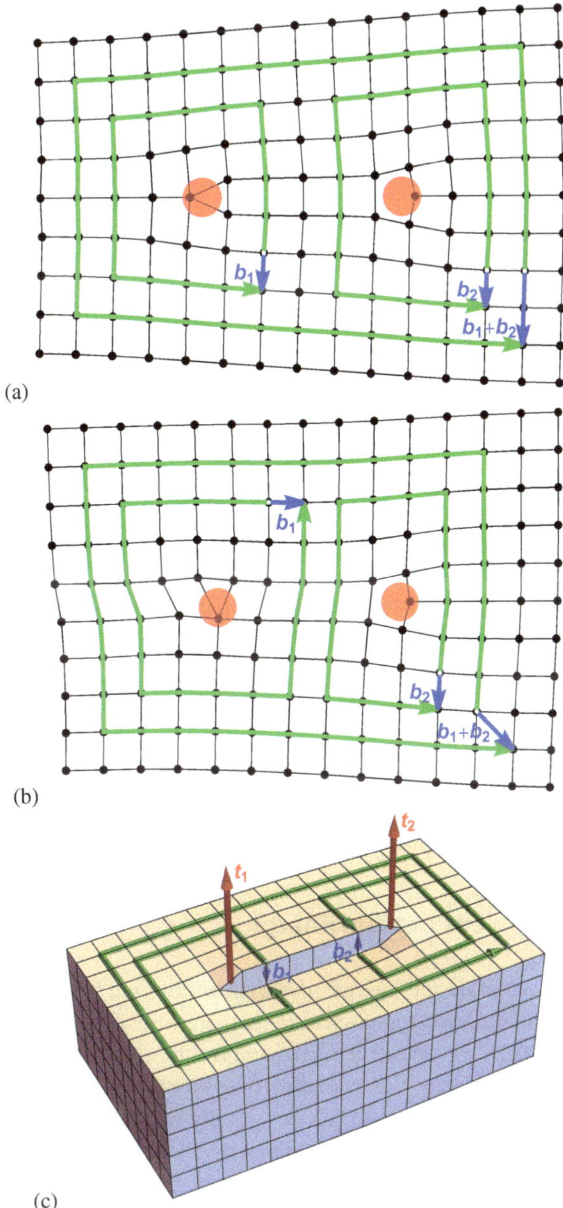

Fig. 9.5 Addition of Burgers vectors. (**a**,**b**) 2D square lattice with two edge dislocations. (**c**) 3D simple cubic lattice with two screw dislocations

Burgers circuit shows that the total Burgers vector is $\boldsymbol{b}_1 + \boldsymbol{b}_2 = a\hat{\boldsymbol{x}} - a\hat{\boldsymbol{y}}$. Likewise, Fig. 9.5c shows a 3D simple cubic lattice with screw dislocations of Burgers vector $\boldsymbol{b}_1 = -a\hat{\boldsymbol{z}}$ and $\boldsymbol{b}_2 = a\hat{\boldsymbol{z}}$. The large Burgers circuit shows that the total Burgers vector is zero; i.e. these two dislocations cancel each other.

Because Burgers vectors add as ordinary vectors, it should be possible to express them in terms of some topological charge density. We can apply Eq. (9.3) to a loop in the plane perpendicular to the tangent vector $\hat{\boldsymbol{t}}$, and then transform the loop integral into an area integral using Stokes' theorem, to obtain

$$\begin{aligned} b_i &= \oint_{\text{loop around } R} (\nabla u_i) \cdot d\boldsymbol{l} = \int_{\text{region } R} d\boldsymbol{r} \left[\hat{\boldsymbol{t}} \cdot \nabla \times (\nabla u_i)\right] \\ &= \left[\int_{\text{region } R} d\boldsymbol{r} \left[\epsilon_{jkl} \partial_k (\partial_l u_i)\right]\right] t_j = \left[\int_{\text{region } R} d\boldsymbol{r} \alpha_{ij}\right] t_j . \end{aligned} \tag{9.4}$$

Hence, the Burgers vector $\boldsymbol{b}$ is expressed in terms of the tangent vector $\hat{\boldsymbol{t}}$ and the density tensor

$$\alpha_{ij} = \epsilon_{jkl} \partial_k (\partial_l u_i). \tag{9.5}$$

This is the basic concept of the *Nye dislocation density tensor*.[4] Outside a dislocation, the displacement $u(\boldsymbol{r})$ is a well-behaved function, and its partial derivatives commute, so $\alpha_{ij} = 0$. However, a dislocation is a singular point where $u(\boldsymbol{r})$ is not sufficiently well-behaved, and its partial derivatives do not commute. Hence, α_{ij} has a δ-function spike at a dislocation,[5]

$$\alpha_{ij}(\boldsymbol{r}) = b_i t_j \delta(\boldsymbol{r} - \boldsymbol{r}_{\text{dislocation}}). \tag{9.6}$$

Note that α_{ij} is invariant if both $\boldsymbol{b}$ and $\hat{\boldsymbol{t}}$ change sign. This tensor is analogous to the tensor D_{ij} for 3D nematic disclination lines, discussed in Sect. 8.5, but it does not have the same issue with false positives.

9.2.4 Dislocation Energy

To calculate the energy of a dislocation, we follow a procedure analogous to our calculation for topological defects in the xy model in Sect. 3.1. We must begin by constructing a general expression for the elastic energy in terms of the displacement field $\boldsymbol{u}(\boldsymbol{r})$. The elastic energy does not depend on any *uniform* displacement $\boldsymbol{u}$, as illustrated in Fig. 9.1a. Rather, it only depends on the *strain tensor*, which involves gradients of $\boldsymbol{u}$. In linear elasticity theory, for small displacements, the simplest approximation for the strain tensor is

---

[4] J. F. Nye, "Some Geometrical Relations in Dislocated Crystals," *Acta Metall.* **1**, 153 (1953). See discussion in P. M. Anderson, J. P. Hirth, and J. Lothe, *Theory of Dislocations*, third edition (Cambridge University Press, 2017), Section 1.4.

[5] For further details about this concept in a 2D crystal, see H. S. Seung and D. R. Nelson, "Defects in Flexible Membranes with Crystalline Order," *Phys. Rev. A* **38**, 1005 (1988), Section II. In a 2D crystal, $\hat{\boldsymbol{t}}$ can always be defined as $\hat{z}$, so it can be omitted from the theory.

$$e_{ij} = \frac{1}{2}(\partial_i u_j + \partial_j u_i). \tag{9.7}$$

A simple model for the elastic energy is then

$$F = \int d\boldsymbol{r} \left[\frac{1}{2}\lambda e_{ii} e_{jj} + \mu e_{ij} e_{ij} \right]. \tag{9.8}$$

Here, λ and μ are two elastic constants, known as the *Lamé coefficients*. We have neglected any crystalline anisotropy, which could give additional elastic constants. In a 3D system, λ and μ have dimensions of energy/volume. The parameter μ can be interpreted as the shear modulus. The bulk modulus K, Young's modulus E, and Poisson's ratio ν can all be expressed in terms of λ and μ as

$$K = \frac{3\lambda + 2\mu}{3}, \qquad E = \frac{\mu(3\lambda + 2\mu)}{\lambda + \mu}, \qquad \nu = \frac{\lambda}{2(\lambda + \mu)}. \tag{9.9}$$

In the limit of an incompressible material, we have $\lambda \to \infty$, $K \to \infty$, $E = 3\mu$, and $\nu = 1/2$.

To minimize F over functions $\boldsymbol{u}(\boldsymbol{r})$, we derive the Euler-Lagrange equation. It can be written as

$$\partial_i \sigma_{ij} = 0, \tag{9.10}$$

in terms of the elastic stress tensor

$$\sigma_{ij} = \lambda e_{kk} \delta_{ij} + 2\mu e_{ij}. \tag{9.11}$$

Equation (9.10) is the standard equation for balance of stresses inside an elastic solid. The procedure for solving this equation is too complex to present here, so we refer readers to textbooks on solid elasticity.[6] For a screw dislocation running along the z-axis, with $\hat{\boldsymbol{t}} = \hat{\boldsymbol{z}}$ and $\boldsymbol{b} = b\hat{\boldsymbol{z}}$, the displacement field is

$$\boldsymbol{u}_{\text{screw}} = \frac{\boldsymbol{b}(\phi + \phi_0)}{2\pi}. \tag{9.12}$$

For an edge dislocation running along the z-axis, with $\hat{\boldsymbol{t}} = \hat{\boldsymbol{z}}$ and $\boldsymbol{b}$ in the (x, y) plane, the displacement field is

[6] See P. M. Anderson, J. P. Hirth, and J. Lothe, *Theory of Dislocations*, third edition (Cambridge University Press, 2017), Sections 3.2 and 3.4. In our Eq. (9.13), we have generalized their Eqs. (3.47)–(3.48) to arbitrary $\boldsymbol{b}$ in the (x, y) plane. Also, we have changed the constant of integration. This change is just a constant displacement, independent of position, which is not important.

$$u_{\text{edge}} = \frac{b(\phi + \phi_0)}{2\pi} - \frac{(b \cdot r_\perp)(\hat{t} \times r_\perp)}{4\pi(1-\nu)r_\perp^2} - \frac{\nu(b \times \hat{t})}{4\pi(1-\nu)} + \frac{(1-2\nu)(b \times \hat{t})}{4\pi(1-\nu)} \log\left(\frac{r_\perp}{R}\right). \tag{9.13}$$

In these expressions, we define $\phi = \tan^{-1}(y/x)$ and $r_\perp = (x, y, 0)$. The constant phase ϕ_0 is analogous to the defect phase discussed in Sects. 6.1 and 7.1.4. We have used these expressions (in the incompressible limit with $\nu = 1/2$) to draw Figs. 9.2, 9.3, 9.4, and 9.5.

We can put these expressions for the displacement field back into Eqs. (9.7) and (9.8), and integrate over all space, to calculate the total energy of a dislocation. Just as in Sect. 3.1, the integrals are logarithmically divergent at both short and long length scales. To deal with these divergences, we must cut off the integration. The *minimum* length scale is the dislocation core radius, which is comparable to the lattice constant a. The *maximum* length scale is the system size R in the (x, y) plane. Also, the integrals are proportional to the dislocation length L_z in the z direction, as they should be for line defects. For a screw dislocation, the energy per length becomes

$$\frac{F_{\text{screw}}}{L_z} = \frac{\mu b^2}{4\pi} \log\left(\frac{R}{a}\right). \tag{9.14}$$

Likewise, for an edge dislocation, the energy per length becomes

$$\frac{F_{\text{edge}}}{L_z} = \frac{\mu b^2}{4\pi(1-\nu)} \log\left(\frac{R}{a}\right). \tag{9.15}$$

Note that these energies diverge logarithmically with the system size R. That dependence will be important in Sect. 9.4.2.

9.3 Disclinations

A disclination is a defect that disrupts the orientational order of a crystal. It also has tremendous effects on the translational order.

In many ways, disclinations in crystals are similar to disclinations in 2D polar order or 2D nematic order. Presumably, that is why they are described by the same word *disclination*. However, there are some differences in the visualization and in the energy calculation, as we will see below.

9.3.1 Disclination Structure

For this section, let us begin with a 2D hexagonal lattice. Figure 9.6 shows two examples of disclinations in hexagonal lattices. In Fig. 9.6a, it looks as if someone

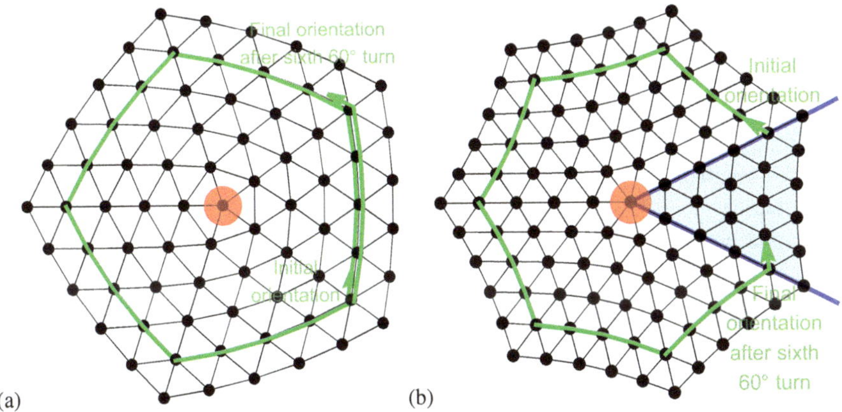

Fig. 9.6 Disclinations in a 2D hexagonal lattice. (**a**) Topological charge $q = +1/6$, with five nearest neighbors around the central site. (**b**) Topological charge $q = -1/6$, with seven nearest neighbors around the central site. In both parts, the green path represents the Burgers circuit, which should close in a perfect lattice. In part (b), the blue wedge represents the extra wedge that has been inserted

has removed a 60° wedge from the lattice, beginning at the red disk and going out to infinity, and then they stretched the remaining lattice to fill up the whole circle. In Fig. 9.6b, it looks as if someone has inserted an extra 60° wedge of particles, which is highlighted in light blue.

Because a perfect hexagonal lattice has sixfold symmetry, the lattice going out from a central point has six identical wedges of 60° each. Hence, the wedge that is removed or inserted must be a multiple of 60°. Afterwards, the remaining lattice fits together perfectly, so that there is no defect line along the removal or insertion. The only defect is a point defect at the red disk.

Disclinations are much more severe defects than dislocations. In a dislocation, only a half-row of particles is removed or added. The missing or extra region remains the same size as we go out to infinity. In a disclination, an entire wedge of particles is removed or added. The missing or extra region grows larger as we go out to infinity.

To characterize a disclination, we can again use a loop construction. As in the previous section on dislocations, we draw a Burgers circuit that *should* close in a perfect crystal, and determine whether it *actually* closes in the crystal with a disclination. In each part of Fig. 9.6, the green path shows a hexagonal Burgers circuit. As we go around the circuit, we move forward by four lattice units, turn counter-clockwise by 60°, and repeat for six sides of hexagon. We can see that the circuit does not close properly; it does not end at the same point where it began. Even worse, we are not even pointing in the same direction as when we began. After six counter-clockwise turns of 60° each, our orientation has rotated counter-clockwise by 60° in Fig. 9.6a, or clockwise by 60° in Fig. 9.6b.

If we draw a larger Burgers circuit around a disclination, the change in orientation remains the same, and the distance from the starting point to the ending point becomes larger. This result contrasts with a Burgers circuit around a dislocation, where there is no change in orientation, and the distance from the starting point to the ending point is always the same.

For a mathematical description of the Burgers circuit around a disclination, we cannot integrate the total displacement, as in Eq. (9.3). Instead, we must integrate the total rotation of the lattice, as in Eq. (1.8). Hence, we define the topological charge of a disclination in a crystal as

$$q = \frac{1}{2\pi} \oint d\theta. \tag{9.16}$$

Here, $\oint d\theta$ means the rotation of the lattice as we move around a Burgers circuit that *should* close in a perfect crystal. In Fig. 9.6a, the total rotation angle is $60° = 2\pi/6$ radians in the counter-clockwise (positive) direction, and hence the disclination has a topological charge of $q = +1/6$. In Fig. 9.6b, the total rotation angle is $60° = 2\pi/6$ radians in the clockwise (negative) direction, and hence the disclination has a topological charge of $q = -1/6$. For any disclination in a hexagonal lattice, the total rotation angle must be a multiple of 60°, so that the remaining lattice will fit together perfectly. Hence, q must be a multiple of $1/6$.

The topological charge of a disclination is related to the *coordination number*, which is defined as the number of nearest neighbors for a site in a crystal. In a perfect 2D hexagonal lattice, each site is surrounded by six nearest neighbors, as indicated by the black lines in our figures. However, for the disclination with $q = +1/6$ in Fig. 9.6a, the central site is surrounded by *five* nearest neighbors. Likewise, for the disclination with $q = -1/6$ in Fig. 9.6b, the central site is surrounded by *seven* nearest neighbors. In general, any positive-charged disclination is associated with a reduced coordination number at the defect point, and any negative-charged disclination is associated with an increased coordination number at the defect point, compared with the coordination number of the perfect crystal.

Very similar considerations apply to disclinations in a 2D square lattice, as illustrated in Fig. 9.7. In Fig. 9.7a, it looks as if someone has removed a 90° wedge from the lattice, and then stretched the remaining lattice to fill up the whole circle. Likewise, in Fig. 9.7b, it looks as if someone has inserted an extra 90° wedge of particles, which is highlighted in light blue. Because a perfect square lattice has fourfold symmetry, the wedge that is removed or inserted must be a multiple of 90°, so that the remaining lattice can fit together perfectly.

In each part of Fig. 9.7, the green loop shows a square Burgers circuit. As we go around the circuit, we move forward by three diagonal units, turn counter-clockwise by 90°, and repeat for four sides of a square. In a perfect crystal, this Burgers circuit should close exactly. For a loop around a disclination, we do not end at the same position where we began, and we are not even pointing in the same direction as when we began. After four counter-clockwise turns of 90° each, our orientation has rotated by 90° in the counter-clockwise (positive) direction in Fig. 9.7a, or by

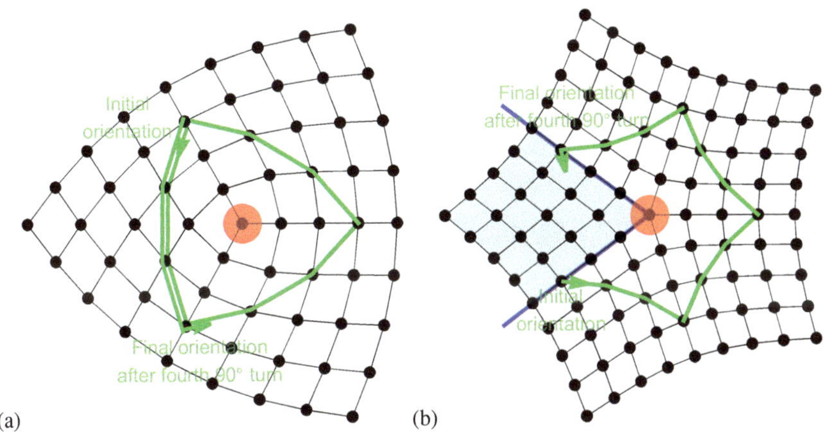

Fig. 9.7 Disclinations in a 2D square lattice. (**a**) Topological charge $q = +1/4$, with three nearest neighbors around the central site. (**b**) Topological charge $q = -1/4$, with five nearest neighbors around the central site. In both parts, the green path represents the Burgers circuit, which should close in a perfect lattice. In part (**b**), the blue wedge represents the extra wedge that has been inserted

90° in the clockwise (negative) direction in Fig. 9.7b. Hence, the disclination in Fig. 9.7a has a topological charge of $q = +1/4$, and the disclination in Fig. 9.7b has a topological charge of $q = -1/4$. For any disclination in a square lattice, q must be a multiple of $1/4$.

Once again, the topological charge of a disclination is related to the coordination number. In a perfect 2D square lattice, each site is surrounded by four nearest neighbors, indicated by black lines in the figures. However, for the disclination with $q = +1/4$ in Fig. 9.7a, the central site is surrounded by *three* nearest neighbors. Likewise, for the disclination with $q = -1/4$ in Fig. 9.7b, the central site is surrounded by *five* nearest neighbors. A positive-charged disclination has a reduced coordination number at the defect point, and a negative-charged disclination has an increased coordination number at the defect point, compared with the coordination number of the perfect crystal.

The concept of a disclination can be extended from a defect point in 2D to a defect line in 3D. This extension is analogous to the way that an edge dislocation was extended from a defect point in 2D to a defect line in 3D, as shown in Fig. 9.4b.

9.3.2 Disclination Energy

The energy of a disclination can be calculated by the same procedure as in Sect. 9.2.4. We use linear elasticity theory, which is based on Eq. (9.7) for the strain tensor and Eq. (9.8) for the elastic energy.[7] The Euler-Lagrange equation is still given by Eq. (9.10), with the stress tensor defined in Eq. (9.11). As in Sect. 9.2.4, the method for solving this equation is too complex to present here, so we refer readers to a relevant article.[8] For a disclination of topological charge q running along the z axis, the displacement field can be expressed in cylindrical coordinates (ρ, ϕ, z) as

$$\boldsymbol{u}_{\text{disclination}} = [q\rho(\phi + \phi_0)]\,\hat{\boldsymbol{\phi}} - \left(\frac{q\rho}{2}\right)\left[1 - \left(\frac{1-2\nu}{1-\nu}\right)\log\left(\frac{\rho}{R}\right)\right]\hat{\boldsymbol{\rho}}, \tag{9.17}$$

where R is the system size in the (x, y) plane.

We can put this expression for the displacement field back into the strain tensor and the elastic energy, and integrate over all space, to calculate the total energy of a disclination. The integral is severely divergent at long length scales in the (x, y) plane. If we cut off the integration at the system size R, we obtain the disclination energy per length

$$\frac{F_{\text{disclination}}}{L_z} = \frac{\pi \mu q^2 R^2}{4(1-\nu)}. \tag{9.18}$$

Note that this energy diverges as R^2 for large system size. That divergence is much stronger than the logarithmic divergence of the dislocation energy calculated in Eqs. (9.14) and (9.15). Hence, an isolated disclination has an extremely high energy, and we should not expect to see isolated disclinations in bulk crystals. Even so, disclinations are still useful for conceptual reasons, as we will see in the following section.

9.4 Relationship Between Dislocations and Disclinations

There is an interesting mathematical relationship between dislocations and disclinations. Here, we present that relationship, and then use it to discuss the 2D melting transition.

---

[7] We recognize that linear elasticity theory may not be a very good approximation here, because the displacements around a disclination are not small. Even so, it gives a reasonable rough estimate of the energy.

[8] H. S. Seung and D. R. Nelson, "Defects in Flexible Membranes with Crystalline Order," *Phys. Rev. A* **38**, 1005 (1988), Section II. The equation for the displacement field is not stated explicitly in this article, but it can be derived using the methods presented there. Note that this article uses 2D elasticity theory. The 2D Poisson ratio σ is related to the 3D Poisson ratio ν by $\sigma = \nu/(1-\nu)$, and the 2D Young's modulus is $K_0 = 2\mu/(1-\nu)$.

9.4.1 Dislocations as Disclination Dipoles

In the theory of electrostatics, we have the concept of an *electric dipole moment* as two opposite electric charges, which are separated by a small distance. In the theory of elasticity, we can construct the analogous concept of an *elastic dipole moment* as two opposite disclinations, which are separated by a small distance. The relationship between dislocations and disclinations is then: *An edge dislocation is a dipole of two opposite disclinations.*

To see this relationship mathematically, we can use linear elasticity theory. Because the theory is linear, any two solutions of the Euler-Lagrange equation can be superposed to make a new solution. Hence, we superpose the solution for a disclination of charge q at position $\boldsymbol{d} = (d_x, d_y)$ with the solution for a disclination of charge $-q$ at position $(0, 0)$,

$$\boldsymbol{u}_{\text{dipole}}(\boldsymbol{r}) = \boldsymbol{u}_{\text{disclination}}(\boldsymbol{r} - \boldsymbol{d}) - \boldsymbol{u}_{\text{disclination}}(\boldsymbol{r}) \tag{9.19}$$

We then expand as a power series for small separation $\boldsymbol{d}$ to obtain

$$\boldsymbol{u}_{\text{dipole}} = -(\boldsymbol{d} \cdot \nabla)\boldsymbol{u}_{\text{disclination}}. \tag{9.20}$$

After some calculation, this result simplifies to

$$\begin{aligned}\boldsymbol{u}_{\text{dipole}} = q(\phi + \phi_0)(\boldsymbol{d} \times \hat{\boldsymbol{t}}) - \frac{q(\boldsymbol{d} \cdot \hat{\boldsymbol{t}} \times \boldsymbol{r}_\perp)(\hat{\boldsymbol{t}} \times \boldsymbol{r}_\perp)}{r_\perp^2} + \frac{q\boldsymbol{d}}{2} \\ - \frac{q(1-2\nu)}{2(1-\nu)}\left[\frac{(\boldsymbol{d} \cdot \boldsymbol{r}_\perp)\boldsymbol{r}_\perp}{r_\perp^2} + \boldsymbol{d}\log\left(\frac{r_\perp}{R}\right)\right].\end{aligned} \tag{9.21}$$

That displacement field is equivalent to $\boldsymbol{u}_{\text{edge}}$ in Eq. (9.13), provided that $\boldsymbol{b}$ and $\boldsymbol{d}$ are related by

$$\boldsymbol{b} = 2\pi q \boldsymbol{d} \times \hat{\boldsymbol{t}}, \qquad \boldsymbol{d} = \frac{\hat{\boldsymbol{t}} \times \boldsymbol{b}}{2\pi q}. \tag{9.22}$$

Hence, the disclination dipole has the same long-range elastic effects as an edge dislocation with Burgers vector $\boldsymbol{b} = 2\pi q \boldsymbol{d} \times \hat{\boldsymbol{t}}$. This Burgers vector is perpendicular to the separation $\boldsymbol{d}$ between the disclinations, and it has magnitude $2\pi q d$.

The relationship between dislocations and disclinations is illustrated in Fig. 9.8. Figure 9.8a shows a hexagonal lattice with two disclinations of topological charge $\pm 1/6$. They are separated by approximately $\boldsymbol{d} \approx 4a\hat{\boldsymbol{x}}$, where a is the lattice constant. The green path represents a hexagonal Burgers circuit around *both* of the disclinations. Around this circuit, the final orientation is consistent with the initial orientation, and the final position displaced from the initial position by a Burgers vector of $\boldsymbol{b} = -4a\hat{\boldsymbol{y}}$. If we draw any other Burgers circuit around *both*

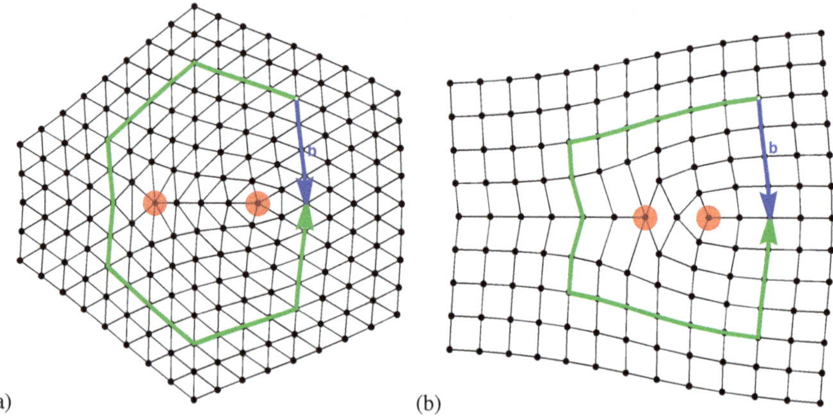

Fig. 9.8 Disclination dipoles. (**a**) Hexagonal lattice with two disclinations of $q = \pm 1/6$. (**b**) Square lattice with two disclinations of $q = \pm 1/4$. In each case, the disclinations can be recognized by their anomalous coordination numbers, and they are highlighted in red. The green paths show Burgers circuits, which should close in perfect crystals. Here, each circuit fails to close by a Burgers vector $\boldsymbol{b}$, showing that the disclination dipole is equivalent to a dislocation

of the disclinations, it will fail to close by the same vector. Hence, the pair of disclinations has a net disclination charge of zero, and it is equivalent to a dislocation of $\boldsymbol{b} = -4a\hat{\mathbf{y}}$. The orientation of $\boldsymbol{b}$ agrees with Eq. (9.22), and the magnitude is approximately correct. It is not exactly correct because it is impossible to identify the disclination positions more precisely than the lattice constant.

Similarly, Fig. 9.8b shows a square lattice with two disclinations of topological charge $\pm 1/4$, separated by approximately $\boldsymbol{d} \approx 2\sqrt{2}a\hat{\mathbf{x}}$. The green path represents a rectangular Burgers circuit around both of the disclinations. The final orientation is consistent with the initial orientation, and the position at the end is displaced from the position at the beginning by $\boldsymbol{b} = -4a\hat{\mathbf{y}}$. Again the orientation of $\boldsymbol{b}$ agrees with Eq. (9.22), and the magnitude is approximately correct, within the lattice resolution.

Using the concept of edge dislocations as disclination dipoles, we can re-examine all of the edge dislocations in Figs. 9.2, 9.3, 9.4, and 9.5. For the hexagonal lattice in Fig. 9.3b, the dislocation includes one site with seven nearest neighbors and another site with five nearest neighbors, separated by approximately $\boldsymbol{d} \approx a\hat{\mathbf{x}}$. Hence, we can regard this dislocation as a dipole of two disclinations, leading to a Burgers vector $\boldsymbol{b} = -a\hat{\mathbf{y}}$.

For the square and simple cubic lattices in the other figures, the analysis is more subtle. In addition to lattice sites, we also need to think about the spaces between lattice sites, known as *plaquettes*. In a perfect 2D square lattice, each plaquette has four nearest-neighbor plaquettes, just as each site has four nearest-neighbor sites. A disclination might be associated with an anomalous coordination number of a site *or* an anomalous coordination number of a plaquette. In Fig. 9.2b, the dislocation includes one plaquette with a coordination number of five, next to one site with a coordination number of three. In that sense, it is a disclination dipole with $\boldsymbol{d} \approx 0.5a\hat{\boldsymbol{x}}$, leading to a Burgers vector $\boldsymbol{b} = -a\hat{\boldsymbol{y}}$. Similarly, each of the edge dislocations in Figs. 9.4b and 9.5 includes one five-coordinated plaquette next to one three-coordinated site, or else one five-coordinated site next to one three-coordinated plaquette. Hence, they can all be regarded as disclination dipoles.

9.4.2 2D Melting and the Hexatic Phase

In Sect. 4.2.2, we saw that the concept of topological defects leads to the Kosterlitz-Thouless theory for the order-disorder transition in the xy model. In this section, we will discuss how the concept of dislocations and disclinations leads to a similar theory for 2D melting, which is called the Kosterlitz-Thouless-Halperin-Nelson-Young (KTHNY) theory.[9]

To understand this theory, let us begin with a 2D crystal at low temperature. The crystal will not have any free disclinations, because the disclination energy of Eq. (9.18) is prohibitively high. It may have a few free edge dislocations, because the dislocation energy of Eq. (9.15) is not that high. However, most dislocations will be tightly bound to other dislocations of opposite Burgers vector, so that they do not disrupt the translational order of the crystal at long length scales.

Now suppose we increase the temperature, and see what happens to the dislocations. As in Sect. 4.2.2, the *free energy* of an isolated edge dislocation involves both energy and entropy,

$$F_{\text{edge at any position}} = F_{\text{edge}} - T S_{\text{positions}}, \tag{9.23}$$

where $S_{\text{positions}} = k_B \log N_{\text{positions}}$ is the positional entropy associated with the number of possible positions for the dislocation. We can estimate that $N_{\text{positions}} \approx (R/a)^2$ (neglecting factors of π or other factors associated with the lattice structure), so that $S_{\text{positions}} = 2k_B \log(R/a)$. We can see that F_{edge} and $S_{\text{positions}}$ depend on system size R in the same way. Hence, we can combine them to obtain

$$F_{\text{edge at any position}} = \left(\frac{\mu b^2}{4\pi(1-\nu)} - 2k_B T\right) \log\left(\frac{R}{a}\right). \tag{9.24}$$

By comparison, the free energy for no dislocation is $F_{\text{no dislocation}} = 0$.

---

[9] J. M. Kosterlitz and D. J. Thouless, "Long Range Order and Metastability in Two Dimensional Solids and Superfluids," *J. Phys. C* **5**, L124 (1972); J. M. Kosterlitz and D. J. Thouless, "Ordering, Metastability and Phase Transitions in Two-Dimensional Systems," *J. Phys. C* **6**, 1181 (1973); B. I. Halperin and D. R. Nelson, "Theory of Two-Dimensional Melting," *Phys. Rev. Lett.* **41**, 121 (1978); D. R. Nelson and B. I. Halperin, "Dislocation-Mediated Melting in Two Dimensions," *Phys. Rev. B* **19**, 2457 (1979); A. P. Young, "Melting and the Vector Coulomb Gas in Two Dimensions," *Phys. Rev. B* **19**, 1855 (1979).

At low temperature, the free energy difference $\Delta F = F_{\text{edge at any position}} - F_{\text{no dislocation}}$ is *positive*, meaning that the energetic cost of forming a free dislocation exceeds the entropic benefit. In that regime, free dislocations are rare. However, when the temperature rises above a critical value, ΔF becomes *negative*, so that the entropic benefit of a free dislocation exceeds the energetic cost. Hence, the bound pairs of dislocations now unbind, and free dislocations proliferate. These dislocations destroy the translational order of the crystal. In other words, the crystal melts at that critical temperature.

Above the critical temperature, the system is no longer in the crystalline phase—but what phase is it in? The free dislocations destroy the translational order of the crystal, but they do not destroy the orientational order. Hence, the system still has orientational order associated with the bonds between neighboring particles. Because of this remaining orientational order, the phase is not an isotropic liquid. Rather, it is a phase with orientational order but not translational order, like a nematic liquid crystal.

In the most common case, the original crystalline phase is hexagonal, because that structure is close-packed. The hexagonal crystal melts into a phase with sixfold orientational order, which is called the *hexatic* phase. Presumably this word is short for *hexagonal nematic*, to emphasize both the sixfold symmetry and the liquid-crystal-like nature of the phase. (In a less common case, the original crystalline phase might be a square lattice. It melts into a phase with fourfold orientational order, which is called the *tetratic* phase.)

Once the system is in the hexatic (or tetratic) phase, it has many free dislocations. As we discussed in Sect. 9.4.1, each dislocation is a dipole of two opposite disclinations, which are bound in a pair. Now we must consider whether the disclinations remain bound together.

Because the hexatic phase has lost its translational order, it is fluid rather than solid, so the shear modulus μ goes to zero. For that reason, there is no longer any energy associated with elastic strain, as in Eq. (9.8). Rather, there is only energy associated with variations in orientational order, similar to Eq. (3.1) for the xy model. Hence, the energy of a free disclination is no longer given by Eq. (9.18), which diverges as R^2. Instead, the disclination energy is similar to Eq. (3.10) for a defect in the xy model, which only diverges as $\log(R/a)$.

Now we can repeat the energy-vs-entropy argument for disclinations rather than dislocations. In the hexatic phase, the disclination energy and entropy both scale as $\log(R/a)$. If the temperature is not too high, the energetic cost of forming a free disclination exceeds the entropic benefit, so free disclinations are rare. However, when the temperature rises about a new critical value (higher than the previous critical temperature), the entropic benefit of a free disclination exceeds the energetic cost. At that point, the dislocations unbind into separate disclinations, and hence free disclinations proliferate. These disclinations destroy the remaining orientational order of the hexatic phase. Above that second critical temperature, the phase is an isotropic liquid.

To summarize this section, the concept of dislocations and disclinations leads to a scenario for a two-stage melting process. First, the system has a transition from a crystalline to a hexatic phase. In this transition, dislocations unbind from each other, and the system loses translational order, but it keeps orientational order. Second, at a higher temperature, the system has a transition from a hexatic to an isotropic liquid phase. In that transition, disclinations unbind from each other, and the system loses orientational order. This two-stage melting scenario has been studied in a range of simulations and experiments.[10]

9.5 Disclinations, Dislocations, and Curvature

So far, we have considered crystalline order in a 2D flat plane, as well as in a 3D bulk system. Now suppose we have crystalline order in a 2D membrane that is free to bend into the third dimension. In that case, disclinations and dislocations are coupled with curvature of the membrane.

To model a flexible membrane, we must consider both the in-plane stretching energy and the out-of-plane bending energy. The in-plane stretching energy is given by Eq. (9.8). This expression is integrated over the 2D membrane, and the coefficients λ and μ are proportional to the membrane thickness t. By contrast, in the bending energy, the coefficients are proportional to t^3. Hence, for a very thin membrane, the stretching energy dominates, and it plays the leading role in determining the shape of the membrane.

To minimize the stretching energy, the membrane must avoid elastic strain e_{ij}. Hence, it must maintain the ideal distances between all of the particles in the crystal, i.e. the ideal bond lengths. If the membrane is flat, as in Figs. 9.2–9.3 and 9.5–9.8, then the bond lengths vary around the defects. Some of the bonds are shorter than the ideal length, and other bonds are longer than the ideal. However, if the membrane buckles into the third dimension, then the bond lengths can be closer to the ideal. For that reason, stretching energy induces curvature around defects.

Figure 9.9a shows the energy-minimizing structure of a positive disclination in a hexagonal lattice on a flexible membrane. It forms a cone with positive Gaussian curvature at the center. In cylindrical coordinates (ρ, ϕ, z), the shape of the cone is described by[11]

$$z = \pm\sqrt{2q}\,\rho, \tag{9.25}$$

for small topological charge q. In this case, we have $q = +1/6$. In the cone, at any radius ρ from the center, the circumference is less than $2\pi\rho$. That reduced circumference is compatible with the reduced number of neighbors around a positive

---

[10] For example, see U. Gasser, C. Eisenmann, G. Maret, and P. Keim, "Melting of Crystals in Two Dimensions," *ChemPhysChem* **11**, 963 (2010).

[11] This theory was developed by D. R. Nelson and L. Peliti, "Fluctuations in Membranes with Crystalline and Hexatic Order," *Journal de Physique* **48**, 1085 (1987).

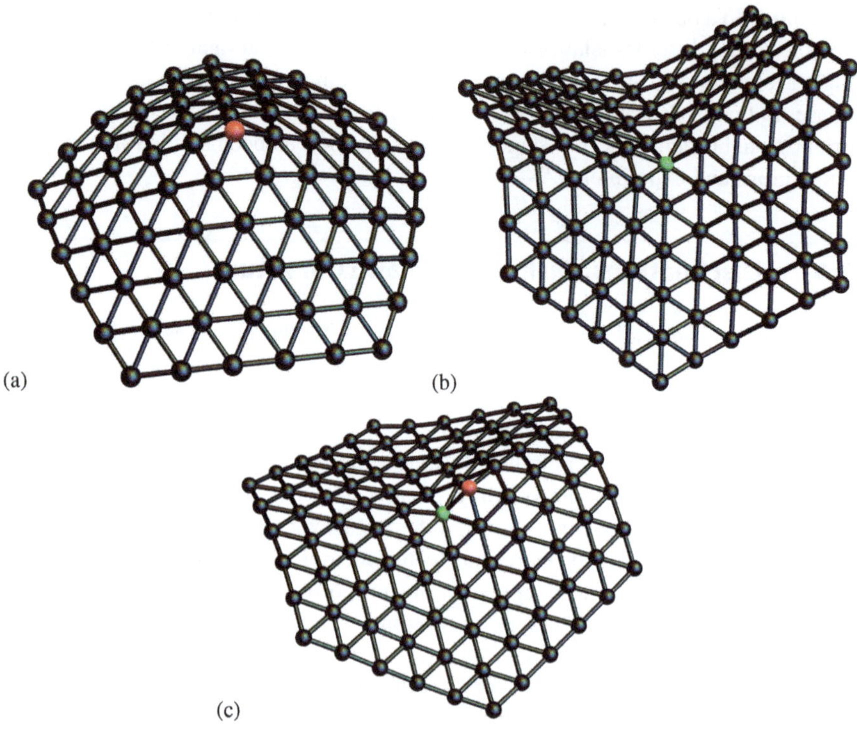

Fig. 9.9 Shapes of flexible membranes with: (**a**) Disclination of topological charge $q = +1/6$, indicated by the red site with five nearest neighbors. (**b**) Disclination of topological charge $q = -1/6$, indicated by the green site with seven nearest neighbors. (**c**) Dislocation (disclination dipole), indicated by the pair of red and green sites. Parts (**a**) and (**b**) are based on analytic minimization of the stretching energy. Part (**c**) is based on data from a numerical minimization, kindly provided by Michael Moshe

disclination. Hence, this buckled disclination has a lower energy than the flat disclination in Fig. 9.6a.

Likewise, Fig. 9.9b shows the energy-minimizing structure of a negative disclination in a hexagonal lattice on a flexible membrane. It forms an anticone with negative Gaussian curvature at the center.[12] Its shape is described by

$$z = \sqrt{\frac{4|q|}{3}}\rho \cos 2(\phi - \phi_0), \tag{9.26}$$

[12] An anticone is similar to a saddle shape, described by $z = \text{const} \cdot \rho^2 \cos 2(\phi - \phi_0)$. However, in an anticone, negative Gaussian curvature is concentrated at the center. In a saddle, negative Gaussian curvature is spread out everywhere.

assuming small $|q|$, for any angle ϕ_0. In this case, we have $q = -1/6$. In the anticone, at a radius ρ from the center, the circumference is greater than $2\pi\rho$. That increased circumference is compatible with the increased number of neighbors around a positive disclination. Hence, this buckled disclination has a lower energy than the flat disclination in Fig. 9.6b.

We can see that coupling of curvature with disclinations in crystalline order is similar to coupling of curvature with disclinations in orientational order, with free energy defined by parallel transport, presented in Sect. 6.3.2. In both cases, positive disclinations favor shapes with positive Gaussian curvature, and negative disclinations favor shapes with negative Gaussian curvature.

As we discussed in Sect. 9.4.1, a dislocation can be regarded as a dipole of two disclinations with opposite topological charges. If a flexible membrane has a dislocation, one might expect the curvature effects of the two opposite charges to cancel each other. That expectation is approximately correct, although the details are more complex because the equations for membrane shape are nonlinear. A detailed numerical study of this problem has been done.[13] Figure 9.9c shows the energy-minimizing structure of a dislocation in a hexagonal lattice. On long length scales, the curvature effects of the disclination dipole partially cancel each other, and the remaining shape is pleated. This buckled dislocation has a lower energy than the flat dislocation in Fig. 9.3b.

The coupling between defects and curvature provides the basic concept for how clothing is designed to fit a human body. A single piece of woven fabric is a crystalline membrane, and its natural shape is flat. By introducing defects, a tailor can change the natural shape to have regions of positive and negative Gaussian curvature, as needed to match the body.

Figure 9.10 shows a dress that has been designed to illustrate this mathematical principle. This dress is made of fabric panels in the shape of polygons. Most of the panels are hexagons (six sides), which correspond to sites in a lattice with six nearest neighbors. They are non-defect sites, and they tend to be flat. However, certain selected panels are pentagons (five sides) or heptagons (seven sides), which correspond to sites with five or seven nearest neighbors. The pentagons are positive disclinations, which generate positive Gaussian curvature. The heptagons are negative disclinations, which generate negative Gaussian curvature. We can see that the bust includes a pentagon, followed by a heptagon, followed by another pentagon. The waist involves a row of heptagons for negative Gaussian curvature, and the hips have a row of pentagons for positive Gaussian curvature. Those defects give the dress a natural shape that fits the shape of the mannequin.

[13] H. S. Seung and D. R. Nelson, "Defects in Flexible Membranes with Crystalline Order," *Phys. Rev. A* **38**, 1005 (1988).

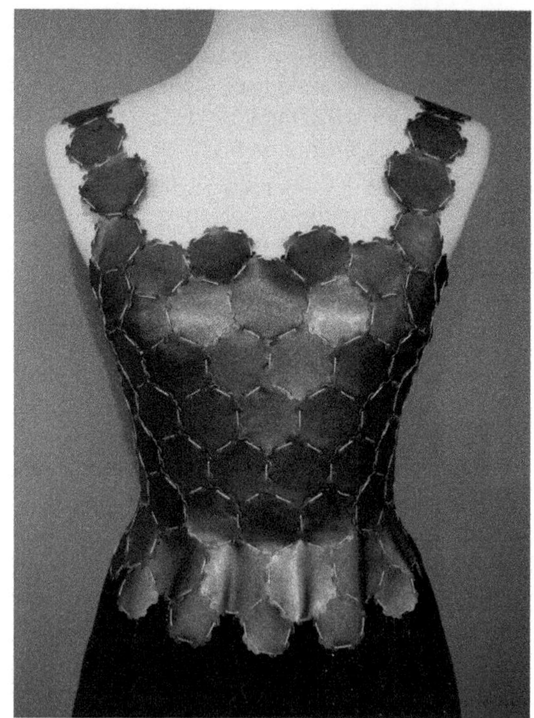

Fig. 9.10 This dress illustrates the relationship between crystalline defects and curvature. The hexagons are non-defect sites, which tend to be flat. Pentagons are positive disclinations, which induce positive Gaussian curvature. Heptagons are negative disclinations, which induce negative Gaussian curvature. The bust has a pentagon, heptagon, and pentagon. The waist has a row of heptagons, and the hips have a row of pentagons. Dress design by Elisabetta Matsumoto, Robin Selinger, and Andrea Schewe, photograph by Andrea Schewe, used by permission

2D Measuring Surface: Hedgehogs, Skyrmions 10

Until now, we have characterized topological features using a 1D measuring surface. For topological defects, we draw a 1D loop *around* a defect, and determine the winding number or Burgers vector that is enclosed by the loop. For topological solitons, we draw a 1D path *through* the soliton, and determine the winding number associated with the soliton.

In this chapter, we generalize to a *2D measuring surface*. We will see that a 2D measuring surface leads to a new type of topological charge, different from the winding number. Based on this 2D topological charge, we can characterize new types of defects and solitons in orientational order.

We first discuss how to use a 2D measuring surface for a system with 3D polar order. After that, we show how the concept must be modified for a system with 3D nematic order.

10.1 3D Polar Order

In this section, we develop the theory for a 2D measuring surface in a system with 3D polar order. Our discussion follows the same steps as in Chap. 1, where we introduced a 1D measuring surface in a system with 2D polar order. Readers might want to refer back to Chap. 1, to see the analogy between these two arguments.

10.1.1 From Winding Number to 2D Topological Charge

To begin, suppose we have 3D polar order inside a disk of radius R. This orientational order is characterized by a unit vector $\hat{\boldsymbol{n}}(\boldsymbol{r}) = (n_x(\boldsymbol{r}), n_y(\boldsymbol{r}), n_z(\boldsymbol{r}))$, which is a function of position $\boldsymbol{r} = (x, y)$. Also suppose there is a boundary condition, which requires $\hat{\boldsymbol{n}} = -\hat{\boldsymbol{z}}$ everywhere on the boundary where $|\boldsymbol{r}| = R$.

J. V. Selinger, *Introduction to Topological Defects and Solitons*, Lecture Notes in Physics 1032, https://doi.org/10.1007/978-3-031-70200-6_10

We will later take the limit of system size $R \to \infty$. The question is: How can $\hat{\boldsymbol{n}}(\boldsymbol{r})$ fill up the entire disk, satisfying three requirements:

1. It is a unit vector, with $|\hat{\boldsymbol{n}}(\boldsymbol{r})| = 1$ for all $\boldsymbol{r}$.
2. It is a smooth, continuous function of $\boldsymbol{r}$.
3. It goes to $-\hat{z}$ along the boundary.

In Fig. 10.1, the left column shows four examples of $\hat{\boldsymbol{n}}(\boldsymbol{r})$ configurations that satisfy all three requirements. We would like to characterize these configurations, in order to determine whether we can continuously transform one configuration into another. For this characterization, we cannot use any 1D measuring path, because we already know that any configuration along a 1D measuring path can escape into the third dimension. Instead, we must consider the entire disk as a 2D measuring surface.

Once again, as in Sect. 8.1, we use *homotopy theory*. For 3D polar order, the set of all possible directors is the unit sphere, shown in the right column of the figure. Any point in position space corresponds to a point in director space (such as the points labeled A, B, C). Any 1D measuring path in position space corresponds to a 1D measuring path in director space (as already presented in Fig. 8.2). Likewise, a 2D measuring surface in position space corresponds to a 2D measuring surface in director space. We want to see how the 2D measuring surface covers the unit sphere. Here are four examples:

- Row (a): In position space, the director is $\hat{\boldsymbol{n}} = -\hat{z}$ everywhere on the disk. In director space, the entire measuring surface corresponds to just a single point, which is the south pole, highlighted in green. This single point covers zero area on the unit sphere.
- Row (b): In position space, as we go outward from the center to the boundary, $\hat{\boldsymbol{n}}$ begins at $-\hat{z}$ (point A), then tips outward (point B), then tips back to $-\hat{z}$ (point C). In director space, the measuring surface covers the green area near the south pole, and then it covers the same green area moving backwards. We can say that it covers the same area in positive and negative ways (which will be defined mathematically below). Hence, the net coverage of the unit sphere is zero.
- Rows (c) and (d): In position space, as we go outward from the center to the boundary, $\hat{\boldsymbol{n}}$ begins at $+\hat{z}$ (point A), then becomes horizontal (point B), then rotates down to $-\hat{z}$ (point C). In director space, the measuring surface covers the entire sphere exactly once, always moving in a positive sense.

From these examples, we can see that the net coverage is a useful way to describe how the measuring surface covers the unit sphere. The net coverage must be an integer (zero, positive, or negative), for any configuration that satisfies the boundary condition. Rows (a) and (b) show two configurations with a net coverage of 0, while rows (c) and (d) show two configurations with a net coverage of +1. It is possible to

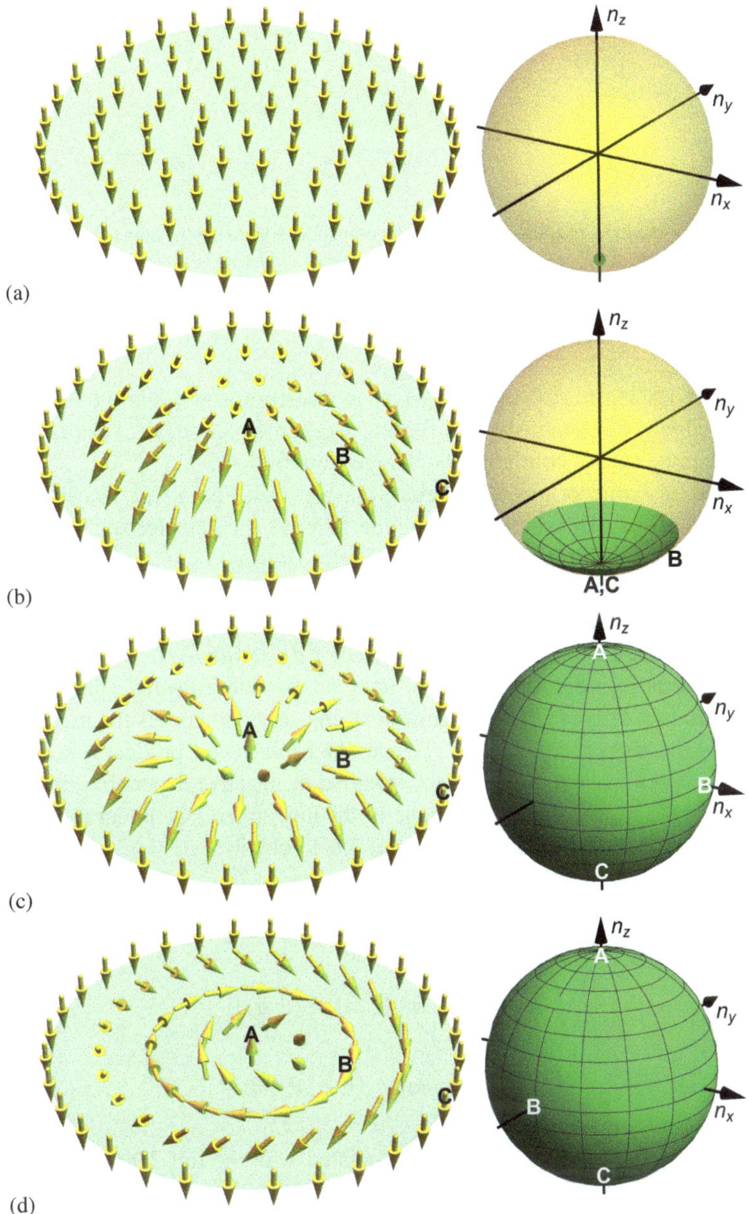

Fig. 10.1 Left column: The yellow arrows represent possible configurations of 3D polar order in position space. All configurations satisfy the boundary condition $\hat{\boldsymbol{n}} = -\hat{z}$ around the edge of the disk. Right column: The yellow sphere represents the set of all possible orientations $\hat{\boldsymbol{n}}$. The green area on the sphere shows which orientations are covered by the configuration on the left. In row (**a**), the only orientation is $\hat{\boldsymbol{n}} = -\hat{z}$, indicated by the green dot at the south pole, so that zero area on the sphere is covered. In row (**b**), the same green area is covered in positive and negative senses, for a net coverage of zero. In rows (**c**) and (**d**), the entire sphere is covered exactly once, in the positive sense

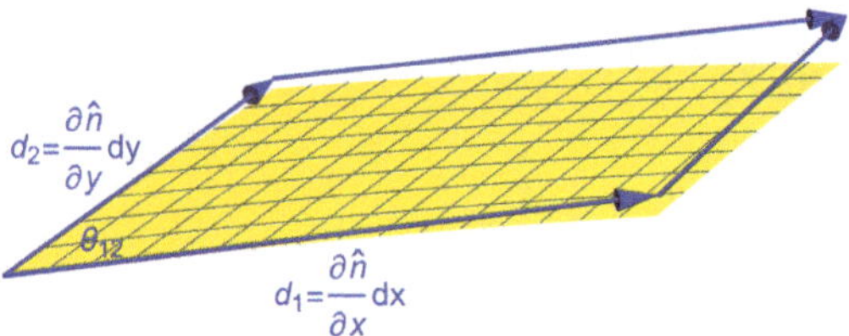

Fig. 10.2 Close-up view of a small region of the unit sphere, showing the area calculation. As we go from x to $x + dx$, and from y to $y + dy$, the director $\hat{\boldsymbol{n}}(x, y)$ covers a parallelogram on the unit sphere. Two edges of the parallelogram are represented by the vectors $\boldsymbol{d}_1$ and $\boldsymbol{d}_2$, and the angle between them is θ_{12}

continuously transform row (a) into (b), or to transform row (c) into (d). However, it is impossible to continuously transform row (a) or (b) into row (c) or (d), because the net coverage cannot continuously change from 0 to +1.

To make this concept more mathematical, we must find a way to calculate the net coverage of the unit sphere. Suppose that $\hat{\boldsymbol{n}}(x, y)$ is parameterized by Cartesian coordinates x and y. If the first coordinate varies from x to $x + dx$, and the second coordinate varies from y to $y + dy$, then $\hat{\boldsymbol{n}}(x, y)$ covers a tiny parallelogram on the unit sphere, as sketched in Fig. 10.2. One edge of the parallelogram is the vector $\boldsymbol{d}_1 = (\partial\hat{\boldsymbol{n}}/\partial x)dx$, and another edge is the vector $\boldsymbol{d}_2 = (\partial\hat{\boldsymbol{n}}/\partial y)dy$. The area of the parallelogram is $|\boldsymbol{d}_1||\boldsymbol{d}_2||\sin\theta_{12}|$. That area can be conveniently calculated from the cross product $\boldsymbol{d}_1 \times \boldsymbol{d}_2$. The magnitude of the cross product is the area, and the direction of the cross product is $\pm\hat{\boldsymbol{n}}$. Hence, the triple vector product $\hat{\boldsymbol{n}} \cdot (\boldsymbol{d}_1 \times \boldsymbol{d}_2)$ is the signed area of the parallelogram, positive or negative depending on the sign of the coverage. To find the net coverage of the unit sphere, we integrate over all x and y, and divide by the total solid angle of 4π, to obtain

$$N = \frac{1}{4\pi}\int dxdy\left[\hat{\boldsymbol{n}} \cdot \left(\frac{\partial\hat{\boldsymbol{n}}}{\partial x} \times \frac{\partial\hat{\boldsymbol{n}}}{\partial y}\right)\right] \tag{10.1a}$$

$$= \frac{1}{8\pi}\int dxdy\left[\epsilon_{3ab}\hat{\boldsymbol{n}} \cdot (\partial_a\hat{\boldsymbol{n}}) \times (\partial_b\hat{\boldsymbol{n}})\right] \tag{10.1b}$$

$$= \frac{1}{8\pi}\int dxdy\left[\epsilon_{3ab}\epsilon_{ijk}n_i(\partial_a n_j)(\partial_b n_k)\right]. \tag{10.1c}$$

This integral is the topological charge associated with the full 2D disk, analogous to the winding number associated with a 1D path. In Fig. 10.1, rows (a) and (b) have $N = 0$, and rows (c) and (d) have $N = +1$. Other configurations could have $N = -1$, or any other positive or negative integer.

We should comment on symmetry and the sign of N. Because Eq. (10.1c) involves *two* Levi-Civita symbols ϵ_{3ab} and ϵ_{ijk}, we can see that N is a proper scalar, not a pseudoscalar. It does not involve any arbitrary choice of right-hand rule vs. left-hand rule. However, it does involve a choice of which direction is $+z$

vs. $-z$. In other words, we are labeling the upper side of the surface as *front* and the lower side as *back*. This labeling tells us which derivative to put first or second in the cross product. In Eq. (10.1a), we write $(\partial\hat{\boldsymbol{n}}/\partial x) \times (\partial\hat{\boldsymbol{n}}/\partial y)$ because $\hat{\boldsymbol{x}} \times \hat{\boldsymbol{y}}$ points from back to front. If we made the opposite choice of which side to label as front or back, then N would change sign. In that respect, the choice of *front* and *back* for a 2D measuring surface is analogous to the choice of *forward* and *backward* for a 1D measuring path.

Mathematical Generalization

It is sometimes necessary to calculate N on a *curved* 2D measuring surface. Also, it is sometimes convenient to calculate N in curvilinear coordinates on a flat 2D measuring surface. For those situations, let us consider a general coordinate system (σ^1, σ^2).

In most of this book, we express tensors in Cartesian coordinates, and hence it is not necessary to use covariant/contravariant tensor notation. However, in this box, because we work in general coordinates, we need covariant/contravariant tensor notation. In that notation, Eq. (10.1b) can be generalized to

$$N = \frac{1}{8\pi} \int d\sigma^1 d\sigma^2 \sqrt{|g|} \left[\epsilon^{ab} \hat{\boldsymbol{n}} \cdot (\partial_a \hat{\boldsymbol{n}}) \times (\partial_b \hat{\boldsymbol{n}}) \right], \tag{10.2}$$

where $g = \det(g_{ij})$ is the determinant of the local covariant metric tensor. Note that the contravariant Levi-Civita tensor components are defined as $\epsilon^{12} = -\epsilon^{21} = +1/\sqrt{|g|}$ and $\epsilon^{11} = \epsilon^{22} = 0$. These factors of $1/\sqrt{|g|}$ cancel the factor of $\sqrt{|g|}$ in the measure of integration.

As an example, if we use polar coordinates (ρ, ϕ) on a flat 2D measuring surface, then N becomes

$$N = \frac{1}{4\pi} \int d\rho d\phi \left[\hat{\boldsymbol{n}} \cdot \left(\frac{\partial \hat{\boldsymbol{n}}}{\partial \rho} \times \frac{\partial \hat{\boldsymbol{n}}}{\partial \phi} \right) \right]. \tag{10.3}$$

Note that the integral is written as $\int d\rho d\phi[\cdots]$, not as $\int \rho d\rho d\phi[\cdots]$, because of the cancellation of the $\sqrt{|g|}$ factors.

In topological notation, the 2D topological charge N is described by the mathematical statement

$$\pi_2(S^2) = \mathbb{Z}. \tag{10.4}$$

Here, the symbol π_2 refers to the second homotopy group, which describes the classification of defects based on 2D measuring surfaces. The symbol S^2 represents the unit sphere, which is the director space for 3D polar order. The symbol $\mathbb{Z}$ is the set of all integers (positive, negative, and zero). Hence, Eq. (10.4) means that the 2D topological charge is an integer.

We emphasize that the 2D topological charge N is different from the 1D topological charge q. They describe distinct topological features of the director configuration. The 1D topological charge q is commonly called the winding number. The 2D topological charge N can be called the *hedgehog charge* or the *skyrmion number*. We will discuss hedgehogs and skyrmions below.

10.1.2 Hedgehogs as Topological Defects

Let us now extend the discussion from the previous section to 3D polar order in a 3D system. Suppose the 3D system is a cylinder with radius R. As in the previous section, we assume that it has boundary conditions such that $\hat{\boldsymbol{n}} = -\hat{z}$ on the curved surface of the cylinder. For this discussion, we do not care about the top and bottom boundaries. One example of an $\hat{\boldsymbol{n}}(\boldsymbol{r})$ configuration satisfying the boundary condition is shown in Fig. 10.3a.

We can calculate the 2D topological charge N by integrating over any 2D measuring surface running across the system. The 2D surface might be horizontal, at fixed z. Alternatively, it might be diagonal, or curved in a complicated way.

If we integrate over any 2D measuring surface that runs *below* the red point, we obtain the 2D topological charge $N = 0$, as in Fig. 10.1a. By contrast, if we

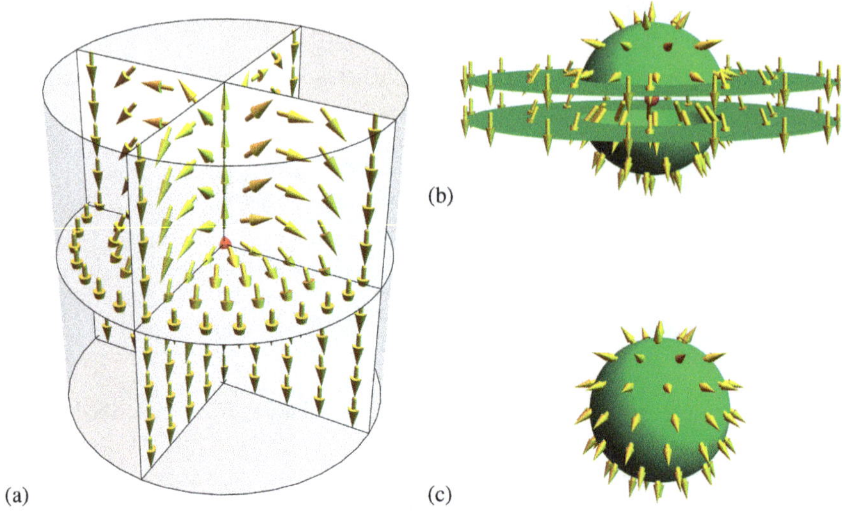

Fig. 10.3 (**a**) Example of a configuration for $\hat{\boldsymbol{n}}(\boldsymbol{r})$ in a 3D system, with a singularity indicated by a red point. (**b**) Composite measuring surface that goes above the singularity (front side oriented upward), and then returns below the singularity (front side oriented downward). (**c**) Closed surface around the singularity (front side oriented outward)

integrate over any 2D measuring surface that runs *above* the red point, with the front side oriented upward, we obtain $N = +1$, as in Fig. 10.1c. If we integrate over any 2D measuring surface that runs exactly through the red point, the integral for N is undefined, because the red point is a singularity where $\hat{\boldsymbol{n}}(\boldsymbol{r})$ is undefined.

Let us now consider the composite measuring surface shown in Fig. 10.3b. This surface runs above the singularity, with the front side oriented upward. It then returns below the singularity, with the front side oriented downward. For the upper part of the surface, the 2D topological charge is $N = +1$, because this part passes above the singularity. For the lower part of the surface, the 2D topological charge is 0, because this part passes below the singularity. Hence, the total 2D topological charge for the composite surface is $N = +1$.

Now we can simplify the surface. In all of the flat regions, our surface goes in both directions, front side oriented upward and downward. These two contributions to the integral cancel each other. Hence, we can remove them from the composite surface. The remaining part is a closed surface, shown in Fig. 10.3c, with the front side oriented outward. The integral over this closed surface is $N = +1$. Likewise, the integral over *any* closed surface that encloses the singularity is $N = +1$. By contrast, the integral over any closed surface that does *not* enclose the singularity is $N = 0$.

Here, we make an intellectual leap analogous to that in Sect. 1.2. Instead of saying that the 2D topological charge is a property of the path, we can say that it is a property of the singularity. In this way of thinking, the singularity is a topological defect. It is not the same type of topological defect that we considered in the previous chapters, with a winding number q. Rather, it is a new type of topological defect, which has a 2D topological charge N. In the literature on liquid crystals, this type of topological defect is called a *hedgehog*.[1] In the literature on magnetism, it is often called a *Bloch point*.

Note that a hedgehog is a defect point in a 3D system. In that respect, it is different from the disclinations studied in previous chapters, which are defect points in a 2D system, or defect lines in a 3D system. This dimensionality is consistent with Eq. (1.10): A hedgehog has $d_{\text{space}} = 3$ and $d_{\text{measure}} = 2$, so that $d_{\text{hedgehog}} = 0$.

We have now defined hedgehogs by an integral over a closed surface inside a system, with no reference to the boundaries. Hence, we can discard the boundary condition, and just think about hedgehogs in an arbitrary geometry, or even in an infinite system. Figure 10.4 shows several examples of hedgehogs in the bulk, away from any boundary. These examples all have 2D topological charge of $N = \pm 1$. Here is an explicit example of how to calculate N using spherical coordinates (r, θ, ϕ) for the hedgehog in Fig. 10.4a:

$$\hat{\boldsymbol{n}} = \frac{\hat{\boldsymbol{x}}x + \hat{\boldsymbol{y}}y + \hat{\boldsymbol{z}}z}{\sqrt{x^2 + y^2 + z^2}} = \hat{\boldsymbol{x}} \sin\theta \cos\phi + \hat{\boldsymbol{y}} \sin\theta \sin\phi + \hat{\boldsymbol{z}} \cos\theta, \tag{10.5}$$

[1] A hedgehog is a small animal that can roll itself into a ball, with its spiny hairs extending outward, like the radial hedgehog defect in Fig. 10.4a. See https://en.wikipedia.org/wiki/Hedgehog.

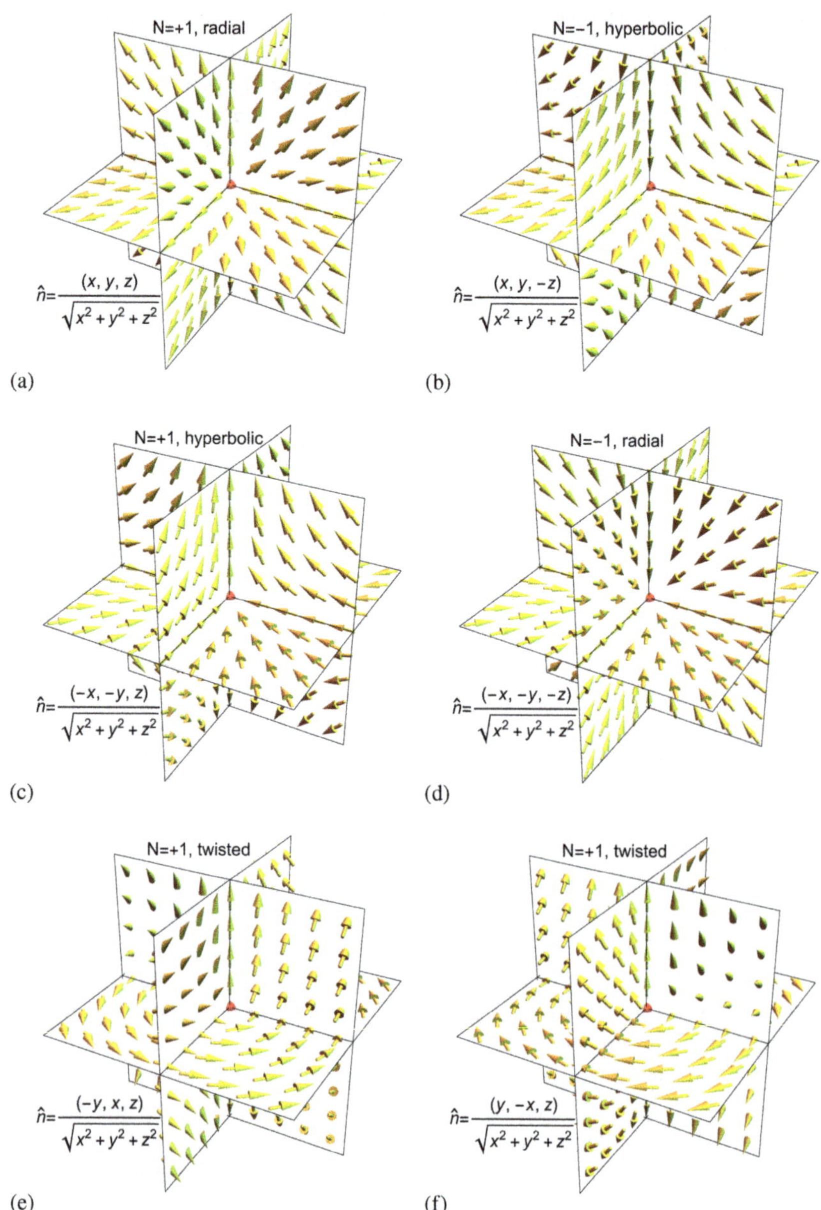

Fig. 10.4 Examples of hedgehogs in a system with 3D polar order. In each case, we indicate the 2D topological charge N, the geometric structure (radial, hyperbolic, or twisted), and the equation for $\hat{\boldsymbol{n}}(x, y, z)$

$$\frac{\partial \hat{\boldsymbol{n}}}{\partial \theta} = \hat{\boldsymbol{x}} \cos\theta \cos\phi + \hat{\boldsymbol{y}} \cos\theta \sin\phi - \hat{\boldsymbol{z}} \sin\theta, \quad \frac{\partial \hat{\boldsymbol{n}}}{\partial \phi} = -\hat{\boldsymbol{x}} \sin\theta \sin\phi + \hat{\boldsymbol{y}} \sin\theta \cos\phi,$$

$$\hat{\boldsymbol{n}} \cdot \left(\frac{\partial \hat{\boldsymbol{n}}}{\partial \theta} \times \frac{\partial \hat{\boldsymbol{n}}}{\partial \phi} \right) = \sin\theta, \quad N = \frac{1}{4\pi} \int_0^{\pi} d\theta \int_0^{2\pi} d\phi \left[\hat{\boldsymbol{n}} \cdot \left(\frac{\partial \hat{\boldsymbol{n}}}{\partial \theta} \times \frac{\partial \hat{\boldsymbol{n}}}{\partial \phi} \right) \right] = 1.$$

In the cross product, we write $\partial\hat{\boldsymbol{n}}/\partial\theta$ first and $\partial\hat{\boldsymbol{n}}/\partial\phi$ second, because $\hat{\boldsymbol{\theta}} \times \hat{\boldsymbol{\phi}}$ points *outward* (back to front) on the measuring surface. In the integral, we write $\int d\theta d\phi[\cdots]$, not $\int \sin\theta d\theta d\phi[\cdots]$, because the triple vector product includes the necessary factors.

In addition to the 2D topological charge N, hedgehogs also have geometric properties. The examples in Fig. 10.4a and d are called *radial hedgehogs*, because the director points radially outward from or inward toward the singularity. The examples in Fig. 10.4b and c are called *hyperbolic hedgehogs*, because the director follows hyperbolic lines around the singularity. The examples in Fig. 10.4e and f are *twisted hedgehogs* with a right- or left-handed structure. Note that the topological charge is not the same as the geometric structure of the hedgehog. A hedgehog with $N = +1$ can be radial, hyperbolic, or twisted, and likewise for $N = -1$. A radial hedgehog can be continuously transformed into a twisted hedgehog by rotating the director by 90° about the z-axis, or into a hyperbolic hedgehog by rotating the director by 180° about the z-axis. A hedgehog with $N = +1$ cannot be continuously transformed into a hedgehog with $N = -1$.

For hedgehogs in a system with 3D polar order, the 2D topological charges combine as addition of ordinary integers. As an example, Fig. 10.5a shows a radial hedgehog with $N = +1$ (on the right) and a hyperbolic hedgehog with $N = +1$ (on the left). These two hedgehogs combine to give a total 2D topological charge of $N = +2$. Far from the hedgehogs, this director field is highly distorted. Similarly, Fig. 10.5b shows a radial hedgehog with $N = +1$ (on the right) and a hyperbolic hedgehog with $N = -1$ (on the left). They combine to give a total of $N = 0$. Far from the hedgehogs, that director field is much less distorted.

Now consider the free energy of a hedgehog. As in Sect. 3.1, our model for the free energy associated with spatial gradients of the director is

$$F = \frac{1}{2} K \int d\boldsymbol{r} |\nabla \hat{\boldsymbol{n}}|^2. \tag{10.6}$$

To minimize this free energy, subject to the constraint that $\hat{\boldsymbol{n}}$ is a unit vector, we derive the Euler-Lagrange equation

$$-K\nabla^2 \hat{\boldsymbol{n}} = \lambda \hat{\boldsymbol{n}}, \tag{10.7}$$

where λ is a Lagrange multiplier to enforce the constraint. All of the hedgehog configurations in Fig. 10.4 satisfy this equation. Hence, we put these configurations back into the free energy density. They all have $|\nabla \hat{\boldsymbol{n}}|^2 = 2/r^2$ in spherical coordinates, with $r^2 = x^2 + y^2 + z^2$. Hence, the total free energy for any of these

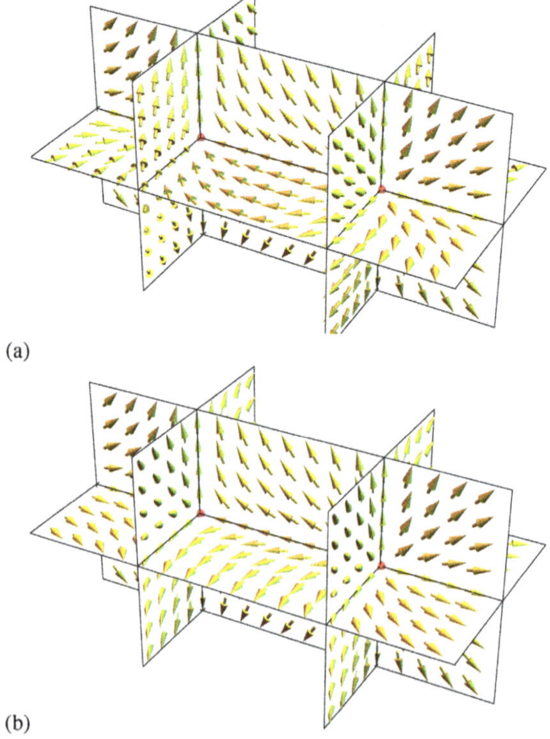

Fig. 10.5 Addition of 2D topological charges for two hedgehogs. (**a**) Radial hedgehog with $N = +1$ (on the right) and hyperbolic hedgehog with $N = +1$ (on the left), giving a total $N = +2$. (**b**) Radial hedgehog with $N = +1$ (on the right) and hyperbolic hedgehog with $N = -1$ (on the left), giving a total $N = 0$

hedgehogs becomes

$$F = \frac{1}{2}K \int d\mathbf{r} \left[\frac{2}{r^2}\right] = \frac{1}{2}K \int r^2 \sin\theta dr d\theta d\phi \left[\frac{2}{r^2}\right] = 2\pi K \int_0^\infty r^2 dr \left[\frac{2}{r^2}\right]. \tag{10.8}$$

This integral can be compared with Eq. (3.9) for the free energy of a topological defect in the xy model. In the current case, for a hedgehog, the integral does *not* diverge at short length scales. Although the free energy *density* diverges as $r \to 0$, the integrated free energy does not diverge there, because the volume around the singularity is so small. However, the integral diverges severely at long length scales. If we cut off the integral at the system radius R, it becomes

$$F = 4\pi K R. \tag{10.9}$$

Hence, the total free energy of a hedgehog is linearly proportional to R. This behavior occurs because the director distortions associated with a hedgehog extend over long distances, far from the singularity. For that reason, isolated hedgehogs have a very high free energy, and are not common in bulk systems.

As an extension of the theory, we can also consider an order parameter with variable magnitude, as in Sect. 5.1. In that case, we write the order parameter as $\boldsymbol{M}(\boldsymbol{r}) = M(\boldsymbol{r})\hat{\boldsymbol{n}}(\boldsymbol{r})$, where M is the magnitude and $\hat{\boldsymbol{n}}$ is the orientation. The free energy then takes the form

$$F = \int d\boldsymbol{r} \left[-\frac{1}{2}a|\boldsymbol{M}|^2 + \frac{1}{4}b|\boldsymbol{M}|^4 + \frac{1}{2}L|\nabla\boldsymbol{M}|^2 \right] \qquad (10.10)$$
$$= \int d\boldsymbol{r} \left[-\frac{1}{2}aM^2 + \frac{1}{4}bM^4 + \frac{1}{2}L|\nabla M|^2 + \frac{1}{2}LM^2|\nabla\hat{\boldsymbol{n}}|^2 \right].$$

The Euler-Lagrange equation for $\hat{\boldsymbol{n}}$ is the same as Eq. (10.7), so the hedgehog solutions still apply. The Euler-Lagrange equation for M becomes

$$-aM + bM^3 - L\nabla^2 M + LM|\nabla\hat{\boldsymbol{n}}|^2 = 0. \qquad (10.11)$$

If we insert $|\nabla\hat{\boldsymbol{n}}|^2 = 2/r^2$ for a hedgehog, and assume that M depends only on r in spherical coordinates, then this equation simplifies to

$$-aM + bM^3 - L\frac{\partial^2 M}{\partial r^2} - \frac{2L}{r}\frac{\partial M}{\partial r} + \frac{2LM}{r^2} = 0. \qquad (10.12)$$

The numerical solution of this equation is quite similar to the corresponding solution in Sect. 5.1, and it gives

$$M(r) \approx M_0(1 - e^{-0.6r/\rho_{\text{core}}}), \qquad (10.13)$$

where $M_0 = \sqrt{a/b}$ and $\rho_{\text{core}} = \sqrt{L/a}$. Hence, the magnitude of orientational order is reduced within a core radius ρ_{core}, and it goes to zero at the center of the hedgehog. By putting this approximate solution back into the free energy and integrating over all space, we obtain the hedgehog free energy relative to a uniform state,

$$\Delta F = 4\pi \underbrace{L \underbrace{\left(\frac{a}{b}\right)}_{M_0^2}}_{K} \left[R - 1.4 \underbrace{\sqrt{\frac{L}{a}}}_{\rho_{\text{core}}} \right]. \qquad (10.14)$$

This free energy is slightly lower than Eq. (10.9), because the magnitude of orientational order is able to relax inside the core region.

10.1.3 Skyrmions as Topological Solitons

Let us now return to a system with 3D polar order as a function of two coordinates x and y. The variation of $\hat{\boldsymbol{n}}(x, y)$ might have a nonzero 2D topological charge, as shown in Fig. 10.1c and d. In most cases, this variation of $\hat{\boldsymbol{n}}(x, y)$ is spread out over the entire size of the system. However, under certain circumstances that will be discussed below, the variation can be concentrated in a small, localized region, while the rest of the system has a uniform background orientation. The localized variation with a nonzero 2D topological charge is called a *skyrmion*. To be more specific, the localized variation of 3D polar order in a 2D system is sometimes called a *baby skyrmion*, in comparison with the original concept of a skyrmion in nuclear physics, which had higher dimensionality.

We will first discuss the structure and topology of skyrmions, and then the free energy and stability.

Figure 10.6 shows two possible structures for skyrmions. In Fig. 10.6a, $\hat{\boldsymbol{n}}(\boldsymbol{r})$ points upward in the center, and then it splays outward in a localized region. Outside that region, it points downward as a uniform background orientation. This structure looks like a modified version of Fig. 10.1c, seen from above, with the splay concentrated over a small length scale. Hence, the splayed region looks like an effective particle with a specific location and a specific size. In this region, $\hat{\boldsymbol{n}}(\boldsymbol{r})$ covers all possible orientations on the unit sphere, in the positive direction, so that the 2D topological charge is $N = +1$. In this context, N is called the *skyrmion number*.

Likewise, in Fig. 10.6b, $\hat{\boldsymbol{n}}(\boldsymbol{r})$ points upward in the center, and then it twists in a right-handed way over a localized region. Outside that region, it points uniformly

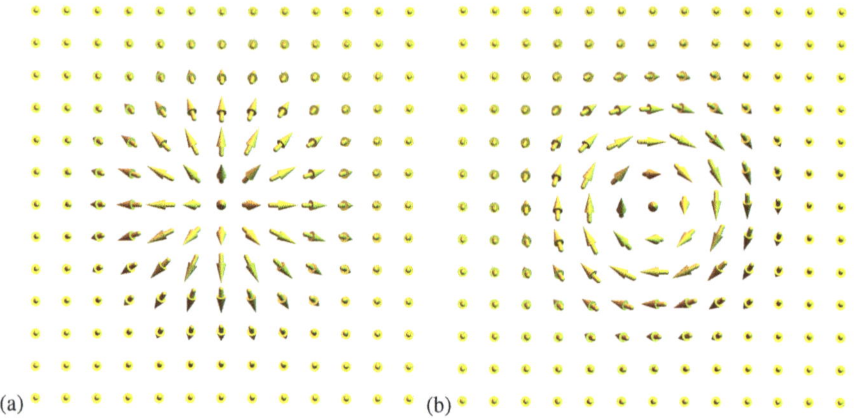

Fig. 10.6 Two possible structures of skyrmions, seen from above. (**a**) Structure with splay, stabilized by surface polarity. (**b**) Structure with twist, stabilized by chirality. In each case, $\hat{\boldsymbol{n}}$ varies in a small, localized region, where it covers all possible orientations on the unit sphere in the positive direction, so that the skyrmion number is $N = +1$. Outside that region, $\hat{\boldsymbol{n}}$ points in a uniform background orientation

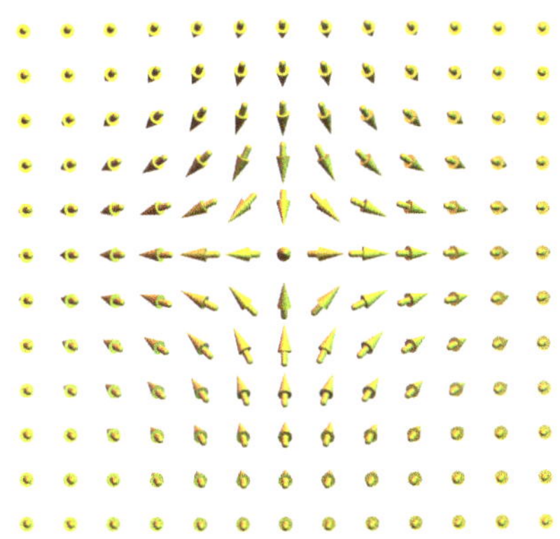

Fig. 10.7 Structure of an antiskyrmion, seen from above. In this case, $\hat{\boldsymbol{n}}(\boldsymbol{r})$ covers all possible orientations on the unit sphere in the negative direction, and hence $N = -1$

downward as a background orientation. This structure looks like a modified version of Fig. 10.1d, seen from above, with the twist concentrated over a small length scale. The twisted region looks like an effective particle with a specific location and a specific size. Again, $\hat{\boldsymbol{n}}(\boldsymbol{r})$ covers all possible orientations on the unit sphere, in the positive direction, so that the skyrmion number is $N = +1$.

For comparison, Fig. 10.7 shows the structure of an antiskyrmion. Like the skyrmions, this antiskyrmion has $\hat{\boldsymbol{n}}(\boldsymbol{r})$ pointing upward in the center, and downward in the uniform background. Unlike the skyrmions, the antiskyrmion is not axisymmetric. Rather, it has regions of right- and left-handed twist and inward and outward splay. In this case, $\hat{\boldsymbol{n}}(\boldsymbol{r})$ covers all possible orientations on the unit sphere, in the negative direction, so that the 2D topological charge is $N = -1$.

In other configurations, which are not shown, the skyrmion number N might be any other integer—positive, negative, or zero.

Based on topology, a skyrmion and an antiskyrmion might annihilate each other, and leave behind a region of the uniform background orientation, with $N = 0$. Conversely, a region of the uniform background orientation might form a skyrmion with $N = +1$ and an antiskyrmion with $N = -1$.

We emphasize that hedgehogs are topological defects, analogous to the defects in Chap. 1. By comparison, skyrmions are topological solitons, analogous to the solitons in Chap. 2. All of the points of comparison between defects and solitons in Table 2.1 still apply to hedgehogs and skyrmions. In particular, hedgehogs have a singularity where $\hat{\boldsymbol{n}}(\boldsymbol{r})$ is undefined. Skyrmions have no singularity; rather, $\hat{\boldsymbol{n}}(\boldsymbol{r})$ is well-defined everywhere. For hedgehogs, the 2D topological charge N is determined by integrating over a 2D measuring surface *around* the hedgehog. For skyrmions, N is determined by integrating over a 2D measuring surface *through* the skyrmion. Hedgehogs are point-like defects in a 3D space, while skyrmions are point-like defects in a 2D plane. Hedgehogs have long-range effects on $\hat{\boldsymbol{n}}(\boldsymbol{r})$, while skyrmions have only short-range effects on $\hat{\boldsymbol{n}}(\boldsymbol{r})$. Away from a skyrmion, $\hat{\boldsymbol{n}}(\boldsymbol{r})$ goes back to its background orientation.

Now consider the free energy of a skyrmion. Under what circumstances is the variation of $\hat{\boldsymbol{n}}(x, y)$ concentrated in a small, localized region? The *Derrick-Hobart theorem* shows that this concentrated variation is unstable if the free energy has the usual elastic form of Eq. (10.6). However, the concentrated variation can occur if the free energy has two additional features:

1. First, the free energy must have some term that gives *orientational alignment* of $\hat{\boldsymbol{n}}$ in the background orientation. This alignment can arise from two possible mechanisms:
   (a) The system might have an *applied magnetic field* $\boldsymbol{B} = -B\hat{\boldsymbol{z}}$, which couples linearly to the magnetic moment density $\boldsymbol{\mu} = \mu\hat{\boldsymbol{n}}$ through a term of $-\boldsymbol{\mu}\cdot\boldsymbol{B} = -\mu\hat{\boldsymbol{n}}\cdot\boldsymbol{B}$, and hence favors the alignment $\hat{\boldsymbol{n}} = -\hat{\boldsymbol{z}}$. This mechanism is the same as considered in Sect. 3.3.
   (b) Alternatively, the system might have *lattice anisotropy*, which favors alignment along an easy axis of the crystalline structure. This anisotropy couples to the orientational order through a term of the form $-(\text{const})(\hat{\boldsymbol{z}}\cdot\hat{\boldsymbol{n}})^2$, which favors alignment of $\hat{\boldsymbol{n}} = \pm\hat{\boldsymbol{z}}$. This mechanism is mathematically equivalent to the term considered in Sect. 7.1.2, although its physical origin may be different.

   Magnetic systems commonly have *both* an applied magnetic field and lattice anisotropy.
2. Second, the free energy must have some term that gives a *favorable variation* of $\hat{\boldsymbol{n}}(\boldsymbol{r})$. This favorable variation may take two possible forms:[2]
   (a) Most commonly, some materials have broken inversion symmetry between $\boldsymbol{r}$ and $-\boldsymbol{r}$. In that case, the most favorable configuration of $\hat{\boldsymbol{n}}(\boldsymbol{r})$ has either a right-handed or a left-handed twist. In the context of magnetism, this effect is called the *Dzyaloshinskii-Moriya interaction*, which arises from the *Dresselhaus spin-orbit coupling* in *noncentrosymmetric* materials. In the context of liquid crystals, this effect is called a *favored twist* due to *chirality*. It is the same effect as considered in Sect. 3.5.
   (b) Alternatively, some materials have a top or bottom surface, which breaks reflection symmetry between z and $-z$. In that case, the most favorable configuration of $\hat{\boldsymbol{n}}(\boldsymbol{r})$ has either inward or outward splay. In the context of magnetism, this effect is another type of Dzyaloshinskii-Moriya interaction, which arises from the *Rashba spin-orbit coupling*. In the context of liquid crystals, this effect is called a *favored splay* due to *surface polarity*.

   In principle, a material might have both favored twist and favored splay, but that is not common.

As a specific example, suppose we have a system with orientational alignment provided by an applied magnetic field pointing downward, and favored twist due to

[2] J. Rowland, S. Banerjee, and M. Randeria, "Skyrmions in Chiral Magnets with Rashba and Dresselhaus Spin-Orbit Coupling," *Phys. Rev. B* **93**, 020404 (2016).

chirality. The total free energy is then

$$F = \int d\boldsymbol{r} \left[\frac{1}{2} K |\nabla \hat{\boldsymbol{n}}|^2 + \mu B - \mu \boldsymbol{B} \cdot \hat{\boldsymbol{n}} + D \hat{\boldsymbol{n}} \cdot \nabla \times \hat{\boldsymbol{n}} \right], \tag{10.15}$$

just as in Eq. (3.26). To achieve the favored twist, $\hat{\boldsymbol{n}}(\boldsymbol{r})$ can take the axisymmetric form

$$\hat{\boldsymbol{n}} = \hat{\boldsymbol{z}} \cos\theta(\rho) - \hat{\boldsymbol{\phi}} \sin\theta(\rho), \tag{10.16}$$

in cylindrical coordinates (ρ, ϕ, z). We can then find the function $\theta(\rho)$ that minimizes the free energy. It goes from $\theta = 0$ at the origin to $\theta = \pi$ as $\rho \to \infty$. The resulting skyrmion structure is shown in Fig. 10.6b. The size of the skyrmion gives the best possible compromise among the terms in the free energy.

If a system has favored splay due to surface polarity, the analysis is fairly similar, but now $\hat{\boldsymbol{n}}(\boldsymbol{r})$ has the axisymmetric form

$$\hat{\boldsymbol{n}} = \hat{\boldsymbol{z}} \cos\theta(\rho) + \hat{\boldsymbol{\rho}} \sin\theta(\rho). \tag{10.17}$$

The resulting skyrmion structure is shown in Fig. 10.6a.

Because an antiskyrmion has regions of right- and left-handed twist and inward and outward splay, it is not favored by either chirality or surface polarity. For that reason, antiskyrmions normally do not form.

Now that we have a model for the $\hat{\boldsymbol{n}}(\boldsymbol{r})$ configuration in a skyrmion, we can compare the free energy of a skyrmion with the free energy of the uniform background. The free energy includes the usual elastic term, the orientational alignment term, and the favorable variation term. In Eq. (10.15), these are the K, μB, and D terms, respectively. For a uniform configuration of $\hat{\boldsymbol{n}}(\boldsymbol{r})$ in the background direction, all of these terms are zero. For a skyrmion configuration, the usual elastic term and the orientational alignment term are positive, while the favorable variation term is negative. Combining these three terms, the total free energy of a skyrmion may be either positive or negative, relative to the uniform background.

If the skyrmion free energy is positive, then skyrmions are metastable. The ground state of a system has no skyrmions, but a typical excited state might have a few isolated skyrmions.

By contrast, if the skyrmion free energy is negative, then skyrmions are actually stable. In that case, a system can reduce its free energy by forming *many* skyrmions. These skyrmions interact with each other through a short-range repulsion, analogous to the short-range repulsion between sine-Gordon solitons in Sect. 3.4. Because of this repulsive interaction, they form a *skyrmion lattice*, as shown in Fig. 10.8a. In this lattice, each skyrmion has $N = +1$, and hence the total 2D topological charge is extremely high. It is not a problem to have many skyrmions with the same 2D topological charge, because skyrmions are solitons, and solitons have only short-range effects on $\hat{\boldsymbol{n}}(\boldsymbol{r})$. There is no need for any antiskyrmions. By contrast, it

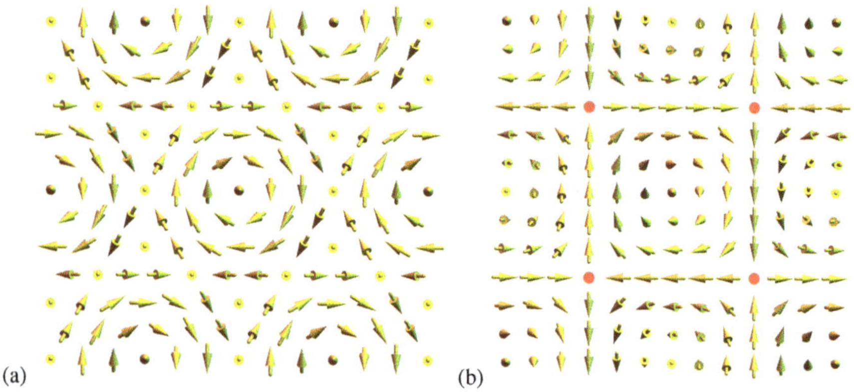

Fig. 10.8 (**a**) Lattice of skyrmions, seen from above. The skyrmion lattice has no antiskyrmions and no defects. (**b**) Lattice of half-skyrmions, also called merons. The half-skyrmion lattice has topological defects with charge $q = -1$, indicated by the red dots

would be impossible to have many hedgehogs with the same 2D topological charge, because hedgehogs are defects, and defects have long-range effects on $\hat{\boldsymbol{n}}(\boldsymbol{r})$. This point was already discussed in the context of Fig. 2.5.

As an alternative to a skyrmion lattice, some systems form a lattice of *half-skyrmions*, also called *merons*. A half-skyrmion has a twist of $\hat{\boldsymbol{n}}(\boldsymbol{r})$ going outward from the center, similar to a full skyrmion, except that it only covers half of the unit sphere. Because it only covers half of the unit sphere, each half-skyrmion must be accompanied by a defect of topological charge $q = -1$. The half-skyrmions and accompanying defects can form a square lattice, illustrated in Fig. 10.8b. This lattice has a half-skyrmion in each square cell, and a defect at each corner where four cells come together, as indicated by the red dots. By contrast, the skyrmion lattice in Fig. 10.8a does not have any defects.

The free energy comparison between the skyrmion lattice and the half-skyrmion lattice is complicated, and it depends on details of the system. In general, the skyrmion lattice is more favorable if the free energy cost of a defect is high. By comparison, the half-skyrmion lattice is more favorable if the free energy cost of a defect is not too high, less than the free energy benefit of a half-skyrmion. In that case, the defects do not escape into the third dimension as in Sect. 8.1, but rather they remain stable.

So far in this section, we have considered skyrmions in a 2D system, where $\hat{\boldsymbol{n}}(\boldsymbol{r})$ depends on $\boldsymbol{r} = (x, y)$. In this 2D system, skyrmions are small, point-like objects. Now, the 2D system can be uniformly extended into 3D, with $\boldsymbol{r} = (x, y, z)$. In a 3D system, skyrmions are extended into lines. An example of a skyrmion line is shown in Fig. 10.9a. If we integrate over any 2D measuring surface passing *through* the skyrmion line, we obtain the skyrmion number $N = 1$. The extension of a skyrmion from a point in 2D to a line in 3D is quite analogous to Sect. 2.2, where we extended a soliton from a point in 1D to a line in 2D or a plane in 3D. The dimensionality

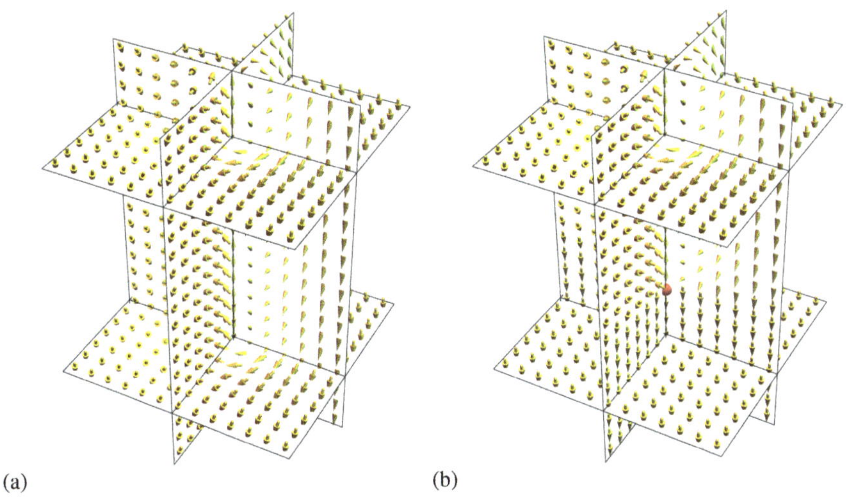

Fig. 10.9 (**a**) Skyrmion line with a concentrated twist of $\hat{\boldsymbol{n}}(\boldsymbol{r})$ running along the z-axis, from the top to the bottom of the figure. (**b**) Skyrmion line running along the z-axis, beginning at the top, and ending at the hedgehog or Bloch point marked in red. In the context of magnetism, structure (**b**) is called a chiral bobber

is consistent with Eq. (2.1): For a skyrmion line in 3D, we have $d_{\text{space}} = 3$ and $d_{\text{measure}} = 2$, and hence $d_{\text{skyrmion}} = 1$.

Like the soliton lines in Fig. 2.3, a skyrmion line can be straight or curved. Just as a soliton line can terminate in a topological defect, a skyrmion line can terminate in a hedgehog or Bloch point. Figure 10.9b shows an example of a skyrmion line ending in a hedgehog, which is indicated in red. If we integrate over any 2D measuring surface passing through the skyrmion line *above* the hedgehog, we obtain $N = +1$. If we integrate over any surface *below* the hedgehog, we obtain $N = 0$. If we integrate over any closed surface *around* the hedgehog, we obtain $N = +1$. Structures like Fig. 10.9b occur experimentally in chiral magnets, with the skyrmion line extending from the surface to a Bloch point in the interior. In that context, the structures are called *chiral bobbers*.

10.2 3D Nematic Order

Let us now consider the theory of hedgehogs and skyrmions in a system with 3D nematic order.

In many ways, the physics of hedgehogs and skyrmions is the same for 3D nematic order as for 3D polar order. To represent 3D nematic order, we can just erase all of the arrowheads from the vectors in Figs. 10.1–10.9. In that case, we are left with cylinders, which are appropriate for a director field in which $\hat{\boldsymbol{n}}$ and $-\hat{\boldsymbol{n}}$ describe the same physical state. For example, Fig. 10.10 shows four examples of hedgehogs in 3D nematic order. These hedgehogs can form with the same geometries as in

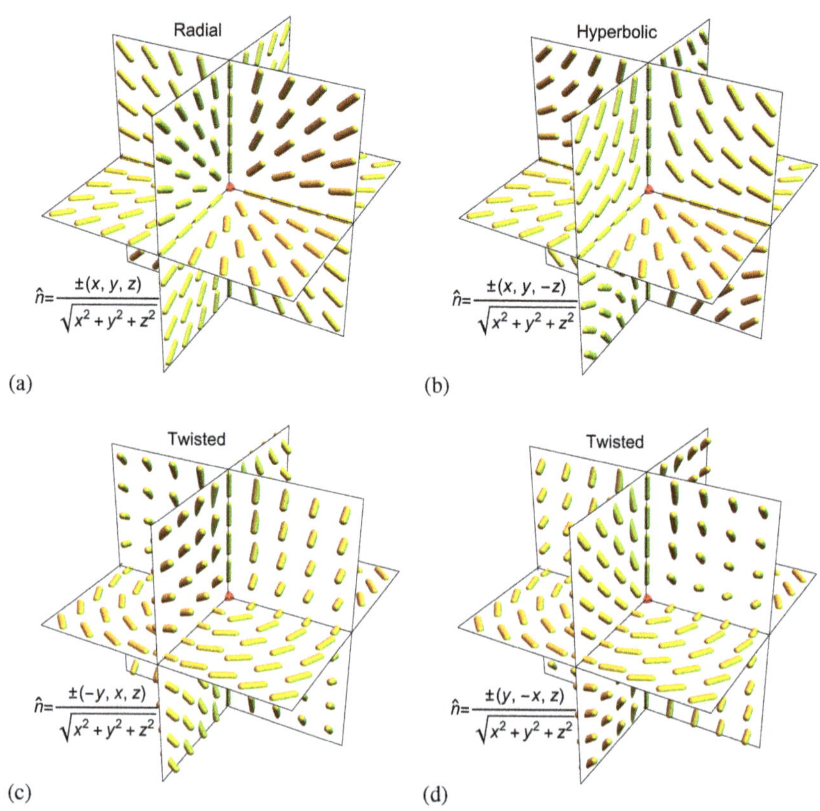

Fig. 10.10 Examples of hedgehogs in a system with 3D nematic order. In each case, the 2D topological charge is $N = \pm 1$. We indicate the geometric structure (radial, hyperbolic, or twisted) and the equation for $\hat{\boldsymbol{n}}(x, y, z)$

Fig. 10.4: radial, hyperbolic, and twisted. Likewise, Fig. 10.11 shows two examples of skyrmions in 3D nematic order. The first structure has splay of the director field, which is stabilized by surface polarity. The second has twist of the director field, which is stabilized by chirality.

Despite these similarities, the theory of hedgehogs and skyrmions is different for 3D nematic order than for 3D polar order in a few important ways. Some of these differences are related to topology, while others are related to free energy. We will outline these differences in the sections below.

10.2.1 What Happens to Topological Charge?

Our first distinction between the theories for nematic and polar order is related to the 2D topological charge N. This distinction is mainly mathematical, rather

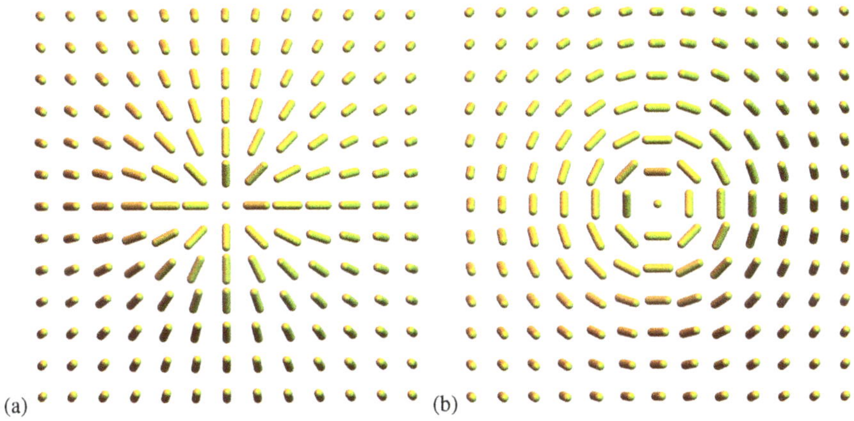

Fig. 10.11 Two possible structures of skyrmions in a system with 3D nematic order, seen from above. (**a**) Structure with splay, stabilized by surface polarity. (**b**) Structure with twist, stabilized by chirality. In each case, the skyrmion number is $N = \pm 1$

than physical. We only put it first because it affects the labeling of hedgehogs and skyrmions.

Let us go back to the definition of N in Eq. (10.1). We can see that this definition involves *three* factors of the director field $\hat{\boldsymbol{n}}$ and its derivatives. The *odd* power of $\hat{\boldsymbol{n}}$ is a problem.[3] In a nematic phase, $+\hat{\boldsymbol{n}}$ and $-\hat{\boldsymbol{n}}$ are two equivalent ways to describe the same physical state. If we substitute $-\hat{\boldsymbol{n}}$ in place of $+\hat{\boldsymbol{n}}$ everywhere in the theory, our description of the physics should not change. However, if we substitute $-\hat{\boldsymbol{n}}$ in place of $+\hat{\boldsymbol{n}}$ in the definition of N, the sign of N will change. Hence, in a nematic phase, $+N$ and $-N$ must be physically equivalent to each other.

As a specific example, consider a radial hedgehog. In a polar phase, there are two different types of radial hedgehogs, with $\hat{\boldsymbol{n}}$ pointing outward or inward. A radial hedgehog pointing outward has $N = +1$ (Fig. 10.4a), and a radial hedgehog pointing inward has $N = -1$ (Fig. 10.4d). By contrast, in a nematic phase, there is no difference between the director pointing outward and inward. Hence, in Fig. 10.10a, we can only label the defect as a radial hedgehog, and we cannot say whether it has $N = +1$ or -1. Likewise, in a polar phase, there are two different types of hyperbolic hedgehogs with $N = +1$ or -1 (Fig. 10.4c and b). In a nematic phase, there is only one type of hyperbolic hedgehog (Fig. 10.10b), and we cannot say whether it has $N = +1$ or -1. In a polar phase, there are right- and left-handed twisted hedgehogs with $N = +1$ (Fig. 10.4e and f), and right- and left-handed

[3] By comparison, this problem does *not* occur for the winding number q. The definition of q in Eq. (1.5) involves *two* factors of $\hat{\boldsymbol{n}}$ and its derivatives.

twisted hedgehogs with $N = -1$ (not shown). In a nematic phase, there are only right- and left-handed twisted hedgehogs (Fig. 10.10c and d).[4]

The same statements are also true for skyrmions. In a polar phase, we can label skyrmions as $N = +1$ and antiskyrmions as $N = -1$. In a nematic phase, the same structures still exist, but we cannot identify the sign of N.

This issue occurs because of the structure of director space, i.e. the ground state manifold. In Sect. 10.1.1, we saw that the definition of N involves the area that is covered in director space, with a positive or negative sign to describe positive or negative coverage. For 3D polar order, that sign can be defined unambiguously, because director space is the unit sphere S^2. This sphere is an *orientable* surface, meaning that we can make a single global choice of which side is outward or inward. However, for 3D nematic order, director space is $S^2/\mathbb{Z}_2$, also written as $\mathbb{R}P^2$, which is the unit hemisphere with the understanding that opposite points on the equator are equivalent. This surface is *non-orientable*, like a Möbius strip or a Klein bottle, meaning that we cannot consistently choose which side is outward or inward. Without that choice, we cannot identify the sign of N.

Readers might ask: Is this issue really a problem? Can't we just treat nematic order as if it were polar order? The answer is: That might or might not work, depending on whether the system contains any disclinations.

If the system does not contain any disclinations, then we can choose one sign of $\hat{\boldsymbol{n}}$ and draw arrowheads on all of the director symbols. This procedure is sometimes called *vectorizing* the director field. We can then analyze properties of nematic order as if it were polar order. In particular, we can calculate N for all of the hedgehogs and skyrmions in the system. If we made the opposite choice for the sign of $\hat{\boldsymbol{n}}$, then we would get the opposite sign of N for all hedgehogs and skyrmions. Hence, all of these signs would be determined by just a single arbitrary choice, which is not a severe problem.

However, *if the system contains any disclinations*, then we cannot consistently choose a sign of $\hat{\boldsymbol{n}}$. In that case, we cannot identify which hedgehogs and skyrmions have positive or negative N, i.e. which can annihilate each other. Rather, we must say that any hedgehog can annihilate any other hedgehog, either immediately or after traveling around a disclination. Likewise, any skyrmion can annihilate any other skyrmion. The topological charges N then combine as

$$0 + 0 = 0, \qquad 0 + 1 = 1, \qquad 1 + 1 = 0. \tag{10.18}$$

These rules can be described as addition of integers, modulo 2.

[4] Some researchers use a convention that radial hedgehogs have $N = +1$, and hyperbolic hedgehogs have $N = -1$, in a nematic phase. For example, see M. Kleman and O. D. Lavrentovich, "Topological Point Defects in Nematic Liquid Crystals," *Philos. Mag.* **86**, 4117 (2006). We prefer not to use this convention, because it can be inconsistent with the labeling in a polar phase.

In a classic review article,[5] David Mermin described the role of disclinations using a remarkable analogy with a *catalyst:* "Note that a nontrivial line defect (of which the 180° disclination is the only type) can catalyze the removal of a point defect [hedgehog] of even index [N].... More generally, a nontrivial disclination can catalyze the elimination of all point defects if the sum of their indices is even, or the transformation of all point defects into a single [$N = 1$] defect if the sum of their indices is odd."

Figure 10.12 shows our visualization of Mermin's catalyst analogy. Initially, in Fig. 10.12a, the system has two identical skyrmions, each apparently with $N = +1$, in the upper-left and upper-right corners. It also has a half-charged (180°) twist disclination in the center, indicated by the red dot. As we go forward in time, in Fig. 10.12b and c, the two skyrmions move downward on *opposite* sides of the disclination. On the left side of the disclination, one skyrmion keeps approximately the same form. On the right side, the other skyrmion continuously changes from a skyrmion to an antiskyrmion, apparently with $N = -1$. In Fig. 10.12d, we have a well-established skyrmion and antiskyrmion on the bottom of the system. In Fig. 10.12e, the skyrmion and antiskyrmion move toward each other and start to merge. Eventually, in Fig. 10.12f, they completely annihilate each other, and leave behind a uniform region of the director field, with $N = 0$. At the end of the process, the disclination is back to its initial state, as a catalyst should be.[6]

The half-charged disclination is a necessary part of this process, which can only happen because the two skyrmions go on opposite sides of the disclination. The disclination can only occur in a system with nematic order, not in a system with polar order. Hence, this example demonstrates that $N = +2$ is topologically equivalent to $N = 0$ in a system with nematic order, but not in a system with polar order.

We should emphasize that this process is a mathematical idealization to demonstrate a topological principle. The process does not normally occur physically. We only present it here to explain why we do not label nematic hedgehogs or skyrmions by $N = +1$ or -1. By contrast, the following points about nematic hedgehogs and skyrmions are more physical.

10.2.2 Hedgehogs and Disclination Loops

For another aspect of defects in 3D nematic order, let us consider the relationship between hedgehogs and disclinations. This relationship is only an issue for 3D nematic order, not for 3D polar order, for a simple reason: Disclinations do not occur in a system with 3D polar order, because they can escape into the third dimension, as shown in Sect. 8.1.

---

[5] N. D. Mermin, "The Topological Theory of Defects in Ordered Media," *Rev. Mod. Phys.* **51**, 591 (1979).

[6] By comparison, if the initial two skyrmions had skyrmion numbers N_1 and N_2, then the second skyrmion would change to $-N_2$, and the final configuration would have skyrmion number $N_1 - N_2$.

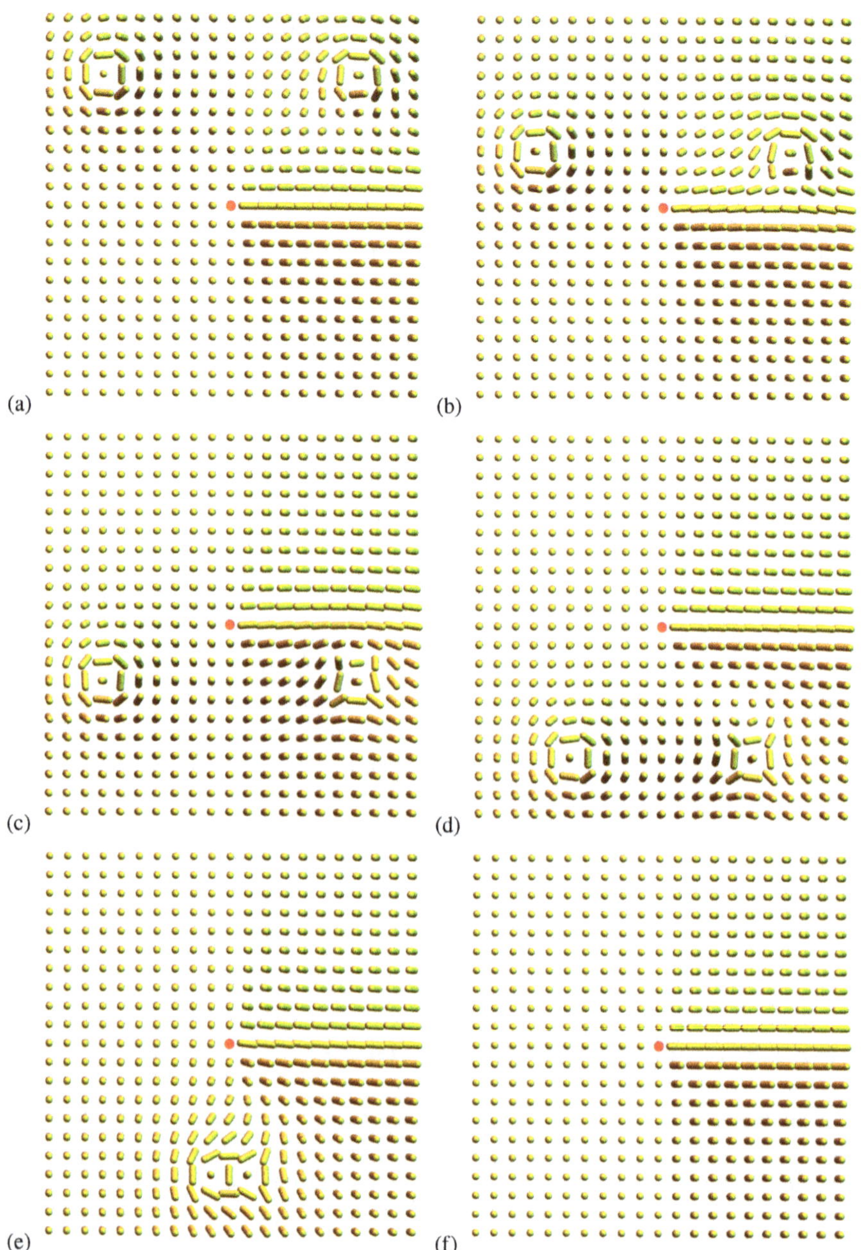

Fig. 10.12 Continuous process in which two nematic skyrmions (each apparently with $N = +1$) move on opposite sides of a disclination and then annihilate each other, leaving a uniform region (with $N = 0$). This process can only occur in a nematic phase, not in a polar phase, because it requires a half-charged disclination

For a system with 3D nematic order, we already know that any disclination has a 1D topological charge, or winding number, of $q = 1/2$. We now ask: Does a disclination *also* have a 2D topological charge N?

To answer that question, recall that N describes what happens if an object is enclosed by a 2D measuring surface. Hence, the first issue is: Can a disclination be enclosed by a 2D measuring surface?

As discussed in Sect. 8.4.1, disclinations are line defects in 3D. A disclination line might go all the way to the boundary of a system, or it might curve into a closed loop. If a disclination line goes all the way to the boundary of a system, then it cannot be enclosed by a 2D measuring surface, and hence N is undefined. However, if the disclination line curves into a closed loop, then it *can* be enclosed by a 2D measuring surface. Thus, N is well-defined for any *closed disclination loop*.

The next question is: For a disclination loop, is N zero or nonzero? The answer is: It depends on the type of loop.

Figure 10.13 shows five examples of disclination loops in a system with 3D nematic order. They are exactly the same five examples that were already illustrated in Fig. 8.5 and presented mathematically in Eq. (8.6), but now visualized in a different way that shows more of the director field. In each case, the red circle represents the disclination loop itself. Let us now consider the director field far outside the loop:

- In Fig. 10.13a, equivalent to Fig. 8.5a, the loop has a $+1/2$ wedge structure everywhere. Far outside the loop, the director field is the same as for the radial hedgehog shown in Fig. 10.10a. Hence, integrating over a 2D measuring surface that encloses this loop is equivalent to integrating over a 2D measuring surface that encloses the radial hedgehog. The resulting 2D topological charge is $N = \pm 1$ (because positive and negative N are equivalent in a system with 3D nematic order, as discussed in the previous section).
- In Fig. 10.13b, equivalent to Fig. 8.5b, the loop has a twisted structure everywhere. Far outside the loop, the director field is the same as for the twisted hedgehog shown in Fig. 10.10c. Integrating over a 2D measuring surface that encloses this loop is equivalent to integrating over a 2D measuring surface that encloses the twisted hedgehog. The resulting 2D topological charge is $N = \pm 1$.
- In Fig. 10.13c, equivalent to Fig. 8.5c, the loop has a $-1/2$ wedge structure everywhere. Far outside the loop, the director field is the same as for the hyperbolic hedgehog shown in Fig. 10.10b. Integrating over a 2D measuring surface that encloses this loop is equivalent to integrating over a 2D measuring surface that encloses the hyperbolic hedgehog. The resulting 2D topological charge is $N = \pm 1$.
- In Fig. 10.13d, equivalent to Fig. 8.5d, the director field is in the (y, z) plane everywhere. As the disclination line curves, the local cross section changes structure: from $+1/2$ to twist to $-1/2$ to twist. In this case, integrating over a 2D measuring surface that encloses the loop gives $N = 0$. That result is reasonable: Because the director field remains in a single plane, it does not cover the unit sphere.

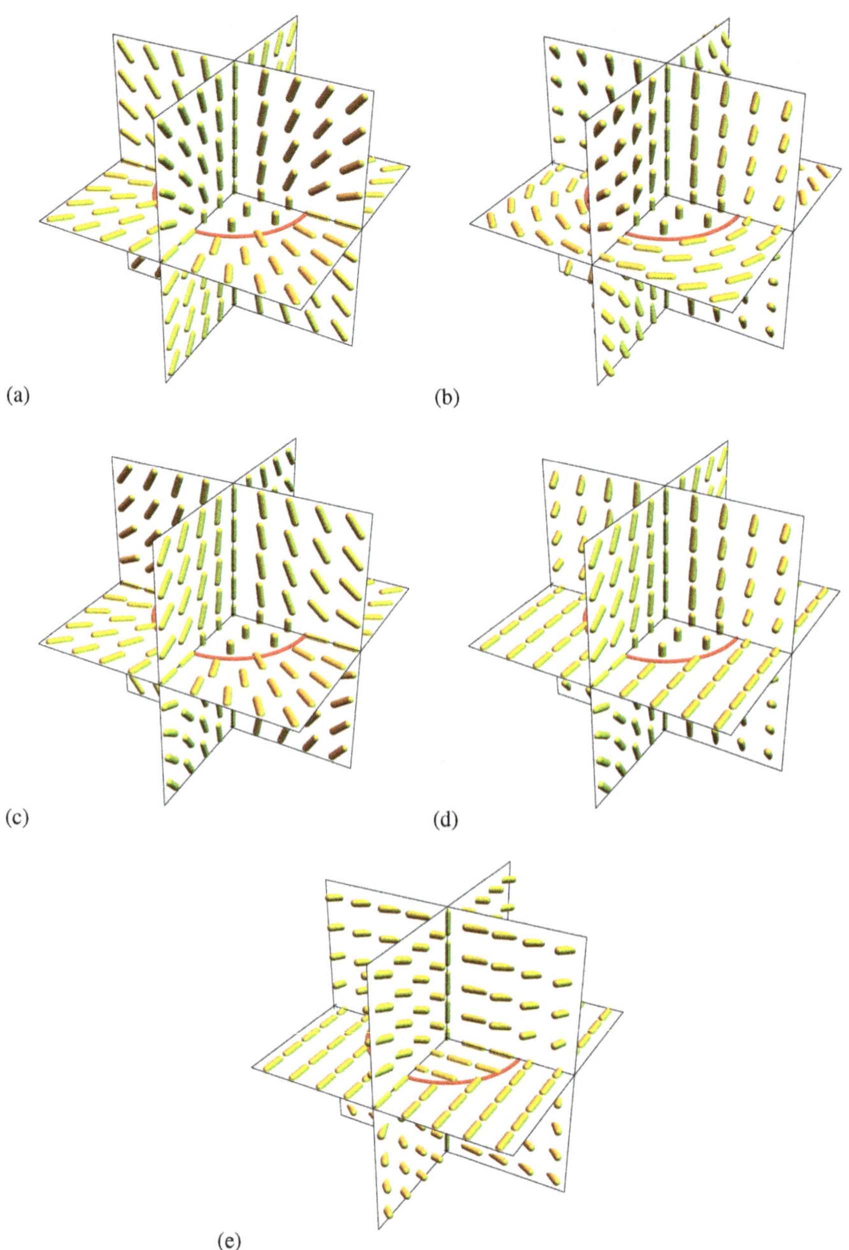

Fig. 10.13 Director field associated with five types of disclination loops. These structures are the same five examples that were already shown in Fig. 8.5. Examples (**a**), (**b**), and (**c**) have a 2D topological charge of $N = \pm 1$, and are topologically equivalent to the hedgehogs in Fig. 10.10a, c, and b, respectively. Examples (**d**) and (**e**) have $N = 0$, and hence they are considered as topologically neutral

- In Fig. 10.13e, equivalent to Fig. 8.5e, the director field is in the (x, y) plane everywhere. The local structure of the disclination is twist everywhere, but it is not the same as Fig. 10.13b. In this case, integrating over a 2D measuring surface that encloses the loop gives $N = 0$, which is again reasonable because the director field remains in a single plane.

This analysis has two important physical consequences for disclination loops in nematic liquid crystals. First, we can see that the disclination loops in Fig. 10.13a–c are topologically equivalent to hedgehogs, meaning that they have the same director field as hedgehogs far outside the loops. The only differences are in a small local region near the loops. As a result, any of these loops can contract into a hedgehog, or conversely, a hedgehog can expand into one of these loops. Either the loop or the hedgehog point might be stable, depending on which has the lower free energy. Indeed, one recent experiment has examined the nanoscale structure of hedgehogs.[7] It finds that radial hedgehogs expand into $+1/2$ disclination loops with a radius of about 40 nm, but hyperbolic hedgehogs remain as point defects. These results do not necessarily apply to all nematic liquid crystals; they might depend on the type of experimental system.

Second, the disclination loops in Fig. 10.13d and e can contract to form a uniform director field, or conversely, they can grow out of a uniform director field, because they have $N = 0$. For that reason, they are sometimes called *topologically neutral* or *topologically trivial*. By contrast, the disclination loops in Fig. 10.13a–c *cannot* contract to form a uniform director field, or grow out of a uniform director field, because they have $N \neq 0$. This distinction is particularly important for *3D active liquid crystals*,[8] which are a 3D generalization of the 2D active liquid crystals discussed in Sect. 7.4. In 2D active liquid crystals, pairs of $+1/2$ and $-1/2$ defects constantly form out of a uniform background, move, and annihilate each other. Similarly, in 3D active liquid crystals, disclination loops constantly form out of a uniform background, move, and contract back into the uniform background. This dynamic behavior only occurs for topologically neutral disclination loops, as in Fig. 10.13d and e. It cannot occur for loops with $N \neq 0$, as in Fig. 10.13a–c.

[7] X. Wang, Y.-K. Kim, E. Bukusoglu, B. Zhang, D. S. Miller, and N. L. Abbott, "Experimental Insights into the Nanostructure of the Cores of Topological Defects in Liquid Crystals," *Phys. Rev. Lett.* **116**, 147801 (2016). This paper labels radial hedgehogs as $+1$, and hyperbolic hedgehogs as -1, following the convention mentioned in footnote 4. We prefer to not to use this convention, as we have discussed.

[8] For an experimental and theoretical study of 3D active liquid crystals, see G. Duclos, R. Adkins, D. Banerjee, M. S. E. Peterson, M. Varghese, I. Kolvin, A. Baskaran, R. A. Pelcovits, T. R. Powers, A. Baskaran, F. Toschi, M. F. Hagan, S. J. Streichan, V. Vitelli, D. A. Beller, and Z. Dogic, "Topological Structure and Dynamics of Three-Dimensional Active Nematics," *Science* **367**, 1120 (2020).

10.2.3 Skyrmions and Torons

We now consider the free energy of skyrmions in 3D nematic order.

As we discussed in Sect. 10.1.3, skyrmions are only stable if the free energy has two specific features, in addition to the usual elastic free energy. First, it must have a term that gives a favorable variation of $\hat{\boldsymbol{n}}(\boldsymbol{r})$. In nematic liquid crystals, this term is normally provided by chirality, which gives a favorable twist. Second, it must have a term that gives an orientational alignment of $\hat{\boldsymbol{n}}$ in the background orientation. In liquid crystals, this term *cannot* be a magnetic field coupling of the form $-\mu\hat{\boldsymbol{n}} \cdot \boldsymbol{B}$, because of the symmetry between $\hat{\boldsymbol{n}}$ and $-\hat{\boldsymbol{n}}$. In principle, it could have the form $-(\Delta\chi/(2\mu_0))(\hat{\boldsymbol{n}} \cdot \boldsymbol{B})^2$, which has the right symmetry, but that situation is not common. The orientational alignment is *usually* provided by *surface anchoring* on the top and bottom of a liquid-crystal cell. If the top and bottom substrates have *homeotropic* (perpendicular) anchoring conditions, then they favor $\hat{\boldsymbol{n}} = \pm\hat{z}$ at the surfaces. Because the elastic free energy favors a uniform director field, the preferred alignment $\hat{\boldsymbol{n}} = \pm\hat{z}$ extends into the interior.

As a mechanism for stabilizing skyrmions, surface anchoring is fairly similar to a magnetic field or crystalline lattice anisotropy, but it is not the same. The main difference is that a magnetic field or crystalline lattice anisotropy acts uniformly, everywhere inside a material. By contrast, surface anchoring acts most strongly on the director field near the top and bottom of the cell, and it only has indirect effects in the interior. For that reason, surface anchoring leads to topological structures with more complex geometric features in 3D, from the top to the bottom of the cell.

Figure 10.14 shows several examples of topological structures that can form. The geometry of these structures depends on the strength of the surface anchoring.

If the surface anchoring is fairly weak, then the director field tends to align in the favored orientation $\hat{\boldsymbol{n}} = \pm\hat{z}$ on most of the top and bottom surfaces, but it does not need to satisfy this anchoring condition everywhere on the surfaces. It can violate the anchoring in small regions of the top and bottom surfaces. As a result, the liquid crystal might form the topological structure shown in Fig. 10.14a. Here, the director field has the preferred orientation everywhere on the top and bottom surfaces except in a small region at the center. The overall structure can be regarded as a skyrmion line from the top to the bottom, including the top and bottom surfaces. The radius of the skyrmion line is small on the top and bottom surfaces, and it expands in the interior of the cell. We can calculate the skyrmion number N by integrating over a 2D measuring surface at any height z. This integral gives $N = \pm 1$ for any value of z, from the top to the bottom.

If the surface anchoring becomes stronger, then the misaligned region on the top and bottom surfaces must become smaller. When the surface anchoring is strong enough, the misaligned region shrinks to a single point on the top, and another single point on the bottom, as shown in Fig. 10.14b. At these two point defects, $\hat{\boldsymbol{n}}$ is undefined. The overall structure can be regarded as a skyrmion line from the top to the bottom, *not* including the top and bottom surfaces. The radius of the skyrmion line is zero on the top and bottom, and it expands in the interior of the cell, reaching

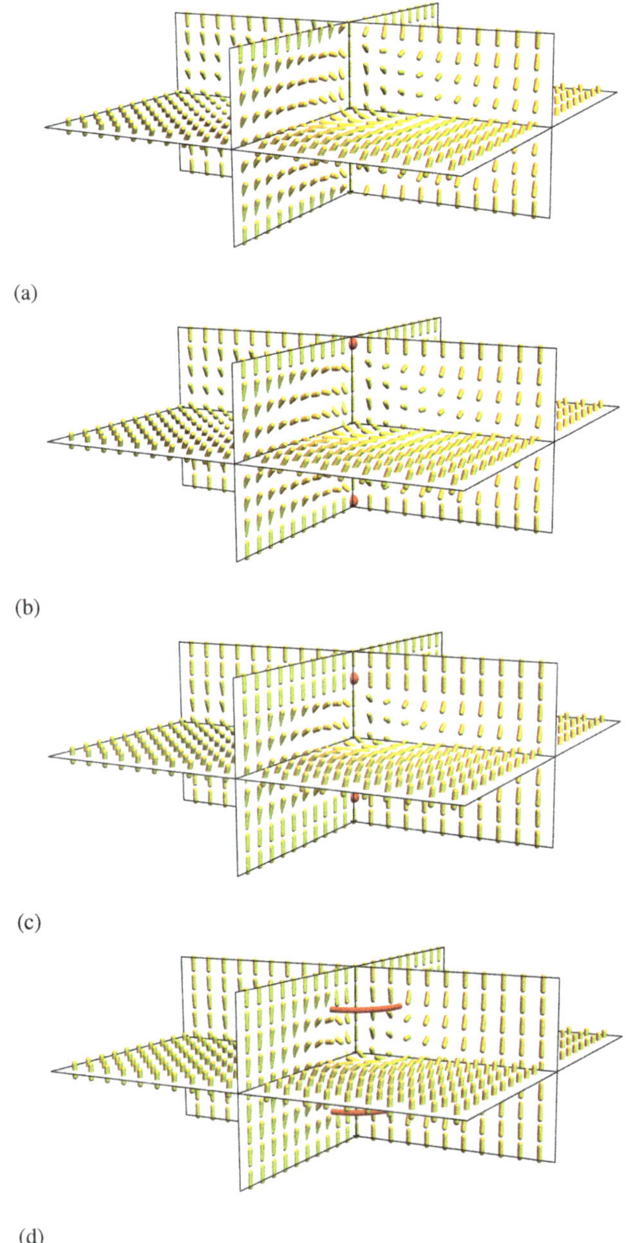

Fig. 10.14 Structure of skyrmions and torons in a cell with 3D nematic order. (**a**) Weak surface anchoring, which is violated in the top center and bottom center. (**b**) Intermediate surface anchoring, which is satisfied everywhere except at single defect points at the top center and bottom center. (**c**) Stronger surface anchoring, which is satisfied everywhere, so that the hedgehog defects are pushed into the interior. (**d**) Variation on the previous structure, with disclination loops instead of point hedgehogs

a maximum radius comparable to the cell thickness. Again, we can calculate the skyrmion number N by integrating over a 2D measuring surface at any height z. This integral gives $N = \pm 1$ for any value of z inside the cell, but it is undefined at the top and bottom surfaces.[9]

If the surface anchoring becomes even stronger, then the point defects are pushed into the interior of the liquid crystal, as shown in Fig. 10.14c. These defect points are now hedgehogs. The overall structure can be regarded as two hedgehogs connected by a skyrmion line. This combined structure is called a *toron*. It is similar to a chiral bobber, as shown in Fig. 10.9b, except that a chiral bobber has a skyrmion line running between the surface and a hedgehog (Bloch point), while a toron has a skyrmion line running between two hedgehogs. Once again, we integrate over a 2D measuring surface to calculate the skyrmion number at any height z. This integral gives $N = 0$ above the upper hedgehog, $N = \pm 1$ between the hedgehogs, and $N = 0$ below the lower hedgehog, and it is undefined for a surface passing through either hedgehog. Each hedgehog is a point where the skyrmion number changes by 1.[10]

As discussed in Sect. 10.2.2, a point hedgehog can expand into a disclination loop, and a disclination loop can contract into a point hedgehog. If the point hedgehogs of Fig. 10.14c expand into disclination loops, we get the structure shown in Fig. 10.14d. This structure is a variation on a toron. It can be regarded as a π-wall soliton in the director field, which is curved into a cylinder that connects the two loops. Like the previous toron, it has $N = \pm 1$ between the two loops, and $N = 0$ above the upper loop and below the lower loop.

The structures of Fig. 10.14 are also called *cholesteric bubbles*, because they form in a cholesteric (chiral nematic) phase, and they look like bubbles when seen from above. Under some circumstances, a cholesteric bubble can become extended in one direction in the (x, y) plane, while it remains small in the orthogonal direction. In that case, it is called a *cholesteric finger*. There are actually several varieties of cholesteric fingers, with and without singularities in the director field $\hat{\boldsymbol{n}}(\boldsymbol{r})$. Cholesteric fingers are beyond the scope of this book, so we will refer interested readers to another book.[11]

[9] For further analysis of this type of structure, see P. J. Ackerman, R. P. Trivedi, B. Senyuk, J. Van De Lagemaat, and I. I. Smalyukh, "Two-Dimensional Skyrmions and Other Solitonic Structures in Confinement-Frustrated Chiral Nematics," *Phys. Rev. E* **90**, 012505 (2014); A. O. Leonov, I. E. Dragunov, U. K. Rößler, and A. N. Bogdanov, "Theory of Skyrmion States in Liquid Crystals," *Phys. Rev. E* **90**, 042502 (2014); S. Afghah and J. V. Selinger, "Theory of Helicoids and Skyrmions in Confined Cholesteric Liquid Crystals," *Phys. Rev. E* **96**, 012708 (2017).

[10] For more information about skyrmions, torons, and other topological structures in confined chiral liquid crystals, readers should see the extensive theoretical and experimental research by the group of Ivan Smalyukh. For example, see P. J. Ackerman and I. I. Smalyukh, Diversity of Knot Solitons in Liquid Crystals Manifested by Linking of Preimages in Torons and Hopfions, Phys. Rev. X 7, 011006 (2017); J.-S. B. Tai and I. I. Smalyukh, "Surface Anchoring as a Control Parameter for Stabilizing Torons, Skyrmions, Twisted Walls, Fingers, and Their Hybrids in Chiral Nematics," *Phys. Rev. E* **101**, 042702 (2020).

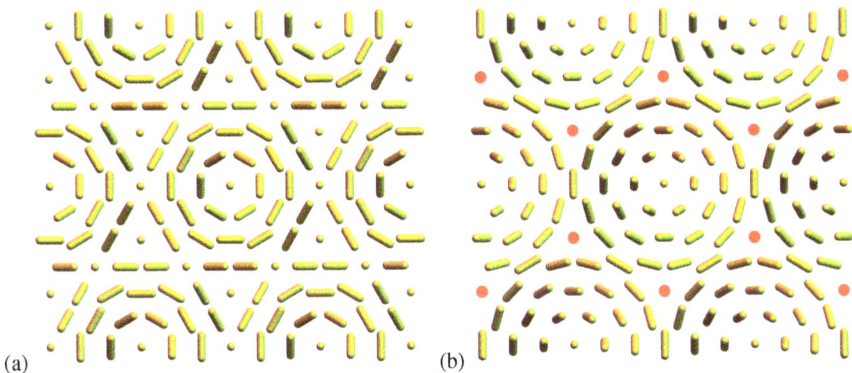

Fig. 10.15 Lattice structures in a system with 3D nematic order. These visualizations show the midplane of a liquid-crystal cell, seen from above. (**a**) Lattice of skyrmions. (**b**) Lattice of half-skyrmions, also called merons. The half-skyrmion lattice has topological defects with charge $q = -1/2$, indicated by red dots

For a skyrmion in nematic order, like a skyrmion in polar order, the total free energy may be either positive or negative, relative to the uniform background. If the skyrmion free energy is positive, then skyrmions are metastable. The ground state has no skyrmions, but a typical excited state might have a few isolated skyrmions. By contrast, if the skyrmion free energy is negative, then skyrmions are actually stable. In that case, a system can reduce its free energy by forming many skyrmions. These skyrmions interact with each other through a short-range repulsion, and hence they may form a skyrmion lattice. One example of a skyrmion lattice is shown in Fig. 10.15a. This figure is exactly the same as Fig. 10.8a except that we have removed all the arrowheads, to represent nematic rather than polar order.

In Fig. 10.8b, we showed that a system with polar order might form a lattice of half-skyrmions, also called merons, as an alternative to a lattice of full skyrmions. This alternative is also possible for a system with nematic order. Once again, a half-skyrmion has a twist of $\hat{\boldsymbol{n}}(\boldsymbol{r})$ going outward from the center, and covering only half of the unit sphere. For the nematic case, each half-skyrmion is accompanied by *two* defects of topological charge $q = -1/2$. The half-skyrmions and accompanying defects can form a hexagonal lattice, as shown in Fig. 10.15b. This lattice has a half-skyrmion in each hexagonal cell, and a defect at each vertex where three cells come together, as indicated by the red dots.[12] The structure of half-skyrmions and defects is similar to a blue phase, which will be discussed in Sect. 12.2.2.

[11] P. Oswald and P. Pieranski, *Nematic and Cholesteric Liquid Crystals: Concepts and Physical Properties Illustrated by Experiments* (CRC Press, 2005).

[12] For a comparison of skyrmion lattices and half-skyrmion lattices in systems with polar and nematic order, see A. Duzgun, J. V. Selinger, and A. Saxena, Comparing Skyrmions and Merons in Chiral Liquid Crystals and Magnets, *Phys. Rev. E* **97**, 062706 (2018).

3D Measuring Surface: Hopfions

11

We can now make one more generalization. In Chaps. 1–9, we characterized topological defects and solitons using a 1D measuring surface, or measuring path. In Chap. 10, we used a 2D measuring surface. In this chapter, we consider a 3D measuring surface, or measuring space. We will see that the 3D measuring space leads to a new type of topological charge and a new type of soliton, although not a new type of defect.

11.1 From Winding Number to 3D Topological Charge

As a first step, we consider an example of a 3D measuring space in a system with 3D polar order. Our discussion follows the same steps as in Chap. 1, for a 1D measuring path with 2D polar order, and in Sect. 10.1, for a 2D measuring surface with 3D polar order.

Suppose we have 3D polar order inside a ball of radius R. This orientational order is characterized by a unit vector $\hat{\boldsymbol{n}}(\boldsymbol{r}) = (n_x(\boldsymbol{r}), n_y(\boldsymbol{r}), n_z(\boldsymbol{r}))$, which is a function of position $\boldsymbol{r} = (x, y, z)$. Also suppose there is a boundary condition, which requires $\hat{\boldsymbol{n}} = \hat{z}$ everywhere on the surface where $|\boldsymbol{r}| = R$. We will later take the limit of system size $R \to \infty$. The question is: How can $\hat{\boldsymbol{n}}(\boldsymbol{r})$ fill up the entire ball, satisfying three requirements:

1. It is a unit vector, with $|\hat{\boldsymbol{n}}(\boldsymbol{r})| = 1$ for all $\boldsymbol{r}$.
2. It is a smooth, continuous function of $\boldsymbol{r}$.
3. It goes to $\hat{z}$ everywhere on the spherical surface of the ball.

In Fig. 11.1, the left column shows three examples of $\hat{\boldsymbol{n}}(\boldsymbol{r})$ configurations that satisfy all three requirements. Row (a) is a uniform configuration with $\hat{\boldsymbol{n}} = \hat{z}$ everywhere in the interior. Rows (b) and (c) are much more complex configurations. These two configurations might seem similar to each other, but they are actually

J. V. Selinger, *Introduction to Topological Defects and Solitons*, Lecture Notes in Physics 1032, https://doi.org/10.1007/978-3-031-70200-6_11

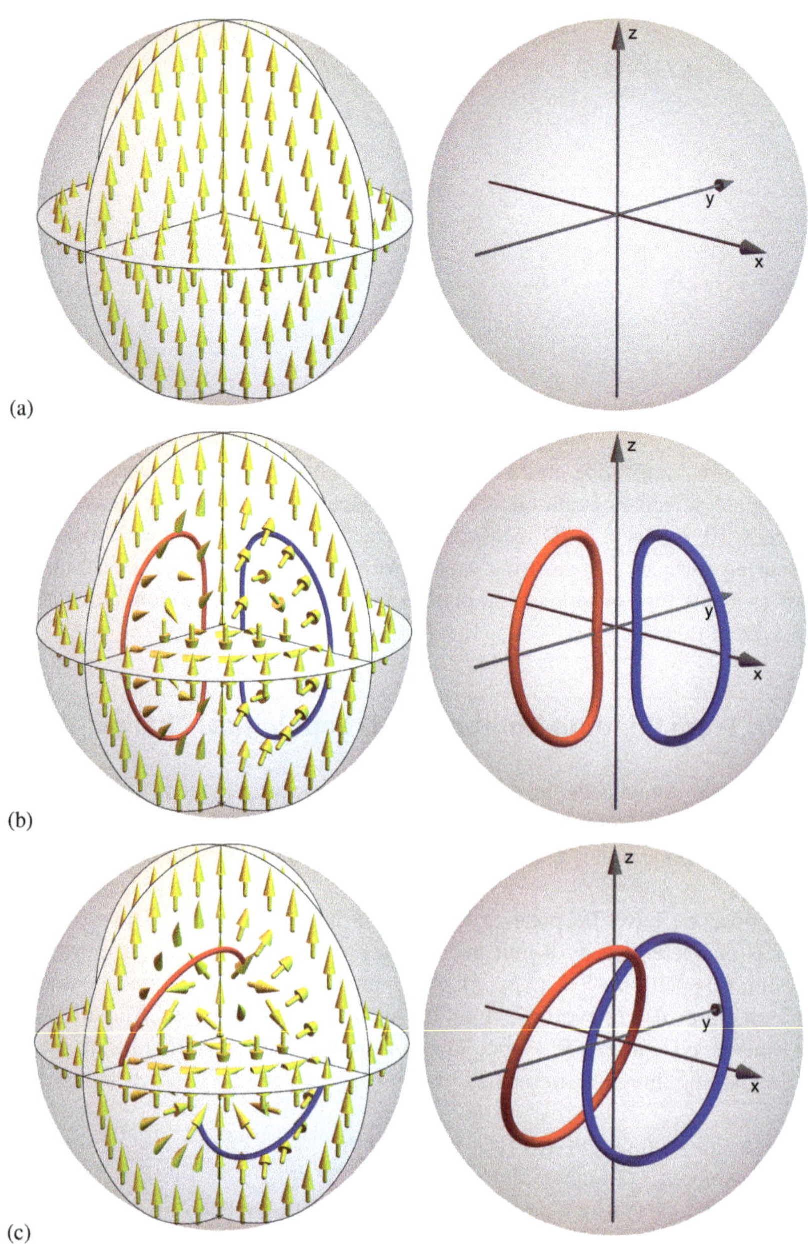

Fig. 11.1 Examples of 3D polar order inside a finite ball, with a boundary condition that requires $\hat{n} = \hat{z}$ everywhere on the surface. The yellow arrows represent $\hat{n}(r)$, and the red and blue curves represent preimages where $\hat{n}(r) = \hat{n}_1 = (\hat{z} + \hat{x})/\sqrt{2}$ or $\hat{n}_2 = (\hat{z} + \hat{y})/\sqrt{2}$, respectively. The right column provides a full view of the preimages in the ball; it is *not* a view of director space. In configurations (**a**) and (**b**), there are no links between preimages, and hence the Hopf index is $Q = 0$. In configuration (**c**), the preimages are linked, and the Hopf index is $Q = 1$. Hence, configurations (**a**) and (**b**) can be continuously transformed into each other, but not into configuration (**c**)

quite different. We claim: Row (b) can be continuously transformed into the uniform configuration of row (a). Row (c) is a different type of configuration, based on the *Hopf fibration* (which will be explained below), and it *cannot* be continuously transformed into a uniform configuration.

How can we check whether this claim is correct? Using a 2D measuring surface is not enough. For any 2D measuring surface that slices through the ball, the 2D topological charge is $N = 0$ for all of the configurations (a), (b), and (c). Instead, we must use the entire ball as a 3D measuring space.

To characterize the configuration in a 3D measuring space, we use the concept of *preimages*. To understand this concept, recall that the function $\hat{\boldsymbol{n}}(\boldsymbol{r})$ creates a correspondence between position space and director space. If we select a specific position $\boldsymbol{r}_{\text{select}}$, it corresponds to a director $\hat{\boldsymbol{n}}(\boldsymbol{r}_{\text{select}})$. We can say that $\hat{\boldsymbol{n}}(\boldsymbol{r}_{\text{select}})$ is the *image* of $\boldsymbol{r}_{\text{select}}$. The same correspondence works in reverse. If we select a specific director $\hat{\boldsymbol{n}}_{\text{select}}$, it corresponds to a *set of positions* $\boldsymbol{r}$, such that $\hat{\boldsymbol{n}}(\boldsymbol{r}) = \hat{\boldsymbol{n}}_{\text{select}}$. We can say that this set of positions is the *preimage* of $\hat{\boldsymbol{n}}_{\text{select}}$. In general, the preimage of any director is a curved line in position space.

Note that the preimages of different directors $\hat{\boldsymbol{n}}_1 \neq \hat{\boldsymbol{n}}_2$ cannot intersect each other. If they intersected at any point $\boldsymbol{r}_{\text{intersect}}$, then we would have $\hat{\boldsymbol{n}}(\boldsymbol{r}_{\text{intersect}}) = \hat{\boldsymbol{n}}_1$ and $\hat{\boldsymbol{n}}(\boldsymbol{r}_{\text{intersect}}) = \hat{\boldsymbol{n}}_2$. Those equations cannot both be true, because $\hat{\boldsymbol{n}}(\boldsymbol{r})$ is a single-valued function.

Let us now choose two unit vectors $\hat{\boldsymbol{n}}_1$ and $\hat{\boldsymbol{n}}_2$, and look at their preimages in Fig. 11.1. We could choose any two arbitrary unit vectors. For a specific example, we choose $\hat{\boldsymbol{n}}_1 = (\hat{z} + \hat{x})/\sqrt{2}$ and $\hat{\boldsymbol{n}}_2 = (\hat{z} + \hat{y})/\sqrt{2}$. In rows (b) and (c), the red curve is the preimage of $\hat{\boldsymbol{n}}_1$, and the blue curve is the preimage of $\hat{\boldsymbol{n}}_2$. The left column presents a partial view of the preimages superimposed on the director field, and the right column presents a full view of the preimages. (In row (a), there are no preimages of $\hat{\boldsymbol{n}}_1$ or $\hat{\boldsymbol{n}}_2$, because $\hat{\boldsymbol{n}}(\boldsymbol{r}) = \hat{z}$ everywhere.)

We can see that the relationship between the red and blue preimages is quite different in rows (b) and (c). In row (b), the preimages are separated from each other, with no linkage between them. By contrast, in row (c), the preimages are linked together, like two links in a metal chain. This distinction is not related to our choice of $\hat{\boldsymbol{n}}_1$ and $\hat{\boldsymbol{n}}_2$. If we choose *any* two unit vectors, the preimages would be unlinked in row (b), but linked in row (c).

This linking shows us which configurations can be transformed into which other configurations. Row (b) can be continuously transformed into the uniform configuration of row (a), because the preimages are unlinked. In this transformation, the red and blue preimages in row (b) would shrink down to nothing. By contrast, row (c) cannot be continuously transformed into the uniform configuration, because there is no way to eliminate the links between the preimages. Elimination of links would require the preimages to cross, and we have already seen that preimages cannot intersect each other. Hence, the preimages in row (c) cannot shrink down to nothing.

The linking number of preimages is an integer called the *Hopf index* Q. It is a topological quantity, because it cannot change under continuous transformations of $\hat{\boldsymbol{n}}(\boldsymbol{r})$ configurations. The configurations in rows (a) and (b) both have $Q = 0$, because they both have no links between preimages. The configuration in row (c) has $Q = 1$. In general, Q can be any integer—positive, negative, or zero. The Hopf index Q serves as a 3D topological charge, analogous to the 1D winding number q and the 2D topological charge N.

The configuration in row (c) can be regarded as a *knot*, which cannot be untied through any continuous transformations of the director field. The Hopf index characterizes this knot.

We have already seen that the 1D winding number q can be calculated as an integral along the 1D measuring path, as shown in Eq. (1.5). We have also seen that the 2D topological charge N can be calculated as a double integral over the 2D measuring surface, as shown in Eq. (10.1). Likewise, the Hopf index Q can be calculated as a triple integral over the entire 3D measuring space.[1] The procedure is: First, construct the vector field $\boldsymbol{B}$ with components

$$B_i = \epsilon_{ijk}\hat{\boldsymbol{n}} \cdot \left[(\partial_j\hat{\boldsymbol{n}}) \times (\partial_k\hat{\boldsymbol{n}})\right]. \tag{11.1}$$

Second, construct the vector field $\boldsymbol{A}$ that satisfies

$$\nabla \times \boldsymbol{A} = \boldsymbol{B}. \tag{11.2}$$

Mathematically, $\boldsymbol{B}$ is analogous to a magnetic field, and $\boldsymbol{A}$ is analogous to a magnetic vector potential. The Hopf index Q can then be determined as

$$Q = -\frac{1}{(8\pi)^2}\int d\boldsymbol{r}(\boldsymbol{B}\cdot\boldsymbol{A}) = -\frac{1}{(8\pi)^2}\int dxdydz(\boldsymbol{B}\cdot\boldsymbol{A}). \tag{11.3}$$

We should note the symmetry of these mathematical objects. Because $\boldsymbol{B}$ involves one cross product and one further Levi-Civita symbol, it is a proper vector. Because $\boldsymbol{A}$ is related to $\boldsymbol{B}$ by a curl, $\boldsymbol{A}$ is a pseudovector. Hence, Q is a pseudoscalar, meaning that it changes sign under reflection. For example, if one $\hat{\boldsymbol{n}}(\boldsymbol{r})$ configuration has $Q = +1$, then the mirror-image configuration has $Q = -1$. In this respect, Q is similar to the 1D winding number q of Eq. (1.5), which is also a pseudoscalar, changing sign under reflection. It is different from the 2D topological charge N of Eq. (10.1), which is a proper scalar that does not change sign under reflection.

Also, let us note what happens under the transformation $\hat{\boldsymbol{n}} \to -\hat{\boldsymbol{n}}$. Clearly, $\boldsymbol{B}$ is odd in $\hat{\boldsymbol{n}}$, so that it changes sign under this transformation. Likewise, $\boldsymbol{A}$ must be odd in $\hat{\boldsymbol{n}}$. As a result, Q is even in $\hat{\boldsymbol{n}}$, so that it does not change sign under this transformation. Once again, Q is similar to the 1D winding number q, which is also even in $\hat{\boldsymbol{n}}$. It is different from the 2D topological charge N, which is odd in $\hat{\boldsymbol{n}}$. Because of this symmetry, the Hopf index Q applies to a phase with 3D nematic

[1] J. H. C. Whitehead, “An Expression of Hopf’s Invariant as an Integral,” *Proc. Natl. Acad. Sci.* **33**, 117 (1947); J. Gladikowski and M. Hellmund, “Static Solitons with Nonzero Hopf Number,” *Phys. Rev. D* **56**, 5194 (1997). See also the helpful website https://hopfion.com/.

order, just as it applies to a phase with 3D polar order. Even in a nematic phase, Q has a well-defined sign; it does not have the sign ambiguity associated with N, which was discussed in Sect. 10.2.1.

To close this section, the following two boxes provide explicit examples of how to construct a director field based on the Hopf fibration, and how to calculate the Hopf index. These boxes are intended for any readers who might want to study these topics in detail. Other readers may wish to skip ahead to the next section.

Construction of Director Field Based on Hopf Fibration

In general, the word *fibration* means a way of breaking space into *fibers*, i.e. straight or curved lines. It is analogous to the word *foliation*, which means a way of breaking space into straight or curved sheets.

The original concept of the Hopf fibration applies to the non-Euclidean curved space S^3, which means the 3D unit sphere embedded in 4D Euclidean space. The 3D unit sphere can be described by the four Cartesian coordinates $(\sigma_1, \sigma_2, \sigma_3, \sigma_4)$, with the constraint that $\sigma_1^2 + \sigma_2^2 + \sigma_3^2 + \sigma_4^2 = 1$. It can also be described by the toroidal coordinate system (θ, ϕ, ω), defined by

$$\sigma_1 = \sin(\theta/2)\cos\phi, \quad \sigma_2 = \sin(\theta/2)\sin\phi,$$
$$\sigma_3 = \cos(\theta/2)\cos\omega, \quad \sigma_4 = \cos(\theta/2)\sin\omega, \tag{11.4}$$

with θ from 0 to π, and ϕ and ω from 0 to 2π. Each surface of constant θ is a torus, and these tori form a foliation of S^3, breaking the space into curved sheets. In the special cases of $\theta = 0$ and $\theta = \pi$, the tori shrink down to great circles.

Within each torus, we can define a series of fibers by the linear relationship between angles ϕ and ω,

$$\phi = \omega + \Phi, \tag{11.5}$$

with Φ constant. Hence, each fiber has a specific value of θ (from 0 to π) and a specific value of Φ (from 0 to 2π). These fibers form the Hopf fibration of S^3. The fibers have several remarkable properties: They are all great circles, all equivalent to each other, and each fiber is linked with each other fiber.

We can now project the tori and fibers from S^3 into ordinary 3D Euclidean space. We use the stereographic projection given by

$$\boldsymbol{r} = (x, y, z) = \frac{(2\sigma_1, 2\sigma_2, 2\sigma_3)}{1+\sigma_4} \tag{11.6}$$
$$= \frac{(2\sin(\theta/2)\cos\phi, 2\sin(\theta/2)\sin\phi, 2\cos(\theta/2)\cos\omega)}{1+\cos(\theta/2)\sin\omega}$$

(continued)

$$= \frac{(2\sin(\theta/2)\cos(\omega+\Phi), 2\sin(\theta/2)\sin(\omega+\Phi), 2\cos(\theta/2)\cos\omega)}{1+\cos(\theta/2)\sin\omega}.$$

By inverting this relationship, we can obtain the angles θ, ϕ, ω, and Φ at any position,

$$\theta = 2\sin^{-1}\left(\frac{4\sqrt{x^2+y^2}}{4+x^2+y^2+z^2}\right), \quad \omega = \tan^{-1}\left(\frac{4-x^2-y^2-z^2}{4z}\right),$$
$$\phi = \tan^{-1}\left(\frac{y}{x}\right), \quad \Phi = \phi - \omega. \tag{11.7}$$

The projection is illustrated in Fig. 11.2. The projected versions of the tori are given by θ constant, with Φ and ω varying from 0 to 2π. The special case of $\theta = \pi$ becomes a circle of radius 2 in the (x, y) plane. The special case of $\theta = 0$ becomes a giant torus, which includes the entire z-axis plus all points at infinity. The projected versions of the fibers are given by θ and Φ constant, with ω varying from 0 to 2π. They wrap around the tori in a helical way, as shown by the black lines in the figure.

To construct a director field, we would like each fiber to be the preimage of some specific unit vector $\hat{\boldsymbol{n}}$. Because each fiber is characterized by specific angles θ and Φ, the unit vector should be characterized by these two angles. Hence, we choose

$$\hat{\boldsymbol{n}} = (\sin\theta\cos\Phi, \sin\theta\sin\Phi, \cos\theta). \tag{11.8}$$

With this choice, the circle of radius 2 in the (x, y) plane becomes the preimage of $\hat{\boldsymbol{n}} = (0, 0, -1)$ at the south pole. The giant torus, including the entire z-axis plus all points at infinity, becomes the preimage of $\hat{\boldsymbol{n}} = (0, 0, +1)$ at the north pole. By combining Eqs. (11.7) and (11.8), we can obtain a general expression for $\hat{\boldsymbol{n}}$ as a function of position. This expression is fairly awkward in Cartesian coordinates, and it is most conveniently written in cylindrical coordinates (ρ, ϕ, z) as

$$\hat{\boldsymbol{n}}(\boldsymbol{r}) = \frac{(32\rho z)\hat{\boldsymbol{\rho}} - 8\rho(4-\rho^2-z^2)\hat{\boldsymbol{\phi}} + (16-24\rho^2+\rho^4+8z^2+z^4+2\rho^2z^2)\hat{\boldsymbol{z}}}{(4+\rho^2+z^2)^2}. \tag{11.9}$$

The director field of Eq. (11.9) has variations that are spread over infinite space, with the boundary condition that $\hat{\boldsymbol{n}} = \hat{\boldsymbol{z}}$ at $\boldsymbol{r} \to \infty$ in any direction. In Fig. 11.1, we want to illustrate director fields inside a ball of radius R, with the boundary condition that $\hat{\boldsymbol{n}} = \hat{\boldsymbol{z}}$ on the spherical surface of the ball. To compress infinite space into a ball, we use the transformation

(continued)

$$r = \frac{2r'}{R - |r'|}, \quad r' = \frac{Rr}{2 + |r|}. \tag{11.10}$$

Figure 11.1c then shows a visualization of $\hat{n}(r')$. In this visualization, the red and blue preimages are two of the fibers from the compressed Hopf fibration.

To be clear, the compressed director field does not necessarily minimize the free energy. It is just a mathematical construction to illustrate a point of topology—it provides a specific example of a director field with linked preimages, and hence a nonzero Hopf index. This mathematical construction would be a good starting point for a numerical minimization of the free energy. If the numerical algorithm makes continuous changes in the director field, then the final relaxed state would also have linked preimages, and hence a nonzero Hopf index.

Fig. 11.2 Stereographic projection of the Hopf fibration into 3D Euclidean space. The tori are surfaces of constant θ, and the fibers (shown in black) are lines of constant θ and Φ. The outer tori and fibers are cut away, so that the inner tori and fibers can be seen. The special case of $\theta = \pi$ corresponds to the innermost torus, which is just a circle of radius 2 in the (x, y) plane. The special case of $\theta = 0$ corresponds to the outermost torus, which includes the entire z-axis plus all points at infinity. Intermediate values of θ correspond to intermediate tori, with fibers wrapping around the tori in a helical way. Each fiber is linked with each other fiber

Calculation of Hopf Index

We have already seen that the Hopf index Q can be represented as an integral over the 3D measuring space, using Eqs. (11.1)–(11.3). These expressions simplify if the director field $\hat{\boldsymbol{n}}$ is parameterized in terms of angles as

$$\begin{aligned}\hat{\boldsymbol{n}} &= (\sin\theta\cos\Phi)\hat{\boldsymbol{x}} + (\sin\theta\sin\Phi)\hat{\boldsymbol{y}} + (\cos\theta)\hat{\boldsymbol{z}} \\ &= [\sin\theta\cos(\Phi-\phi)]\hat{\boldsymbol{\rho}} + [\sin\theta\sin(\Phi-\phi)]\hat{\boldsymbol{\phi}} + (\cos\theta)\hat{\boldsymbol{z}} \\ &= (\sin\theta\cos\omega)\hat{\boldsymbol{\rho}} - (\sin\theta\sin\omega)\hat{\boldsymbol{\phi}} + (\cos\theta)\hat{\boldsymbol{z}},\end{aligned} \tag{11.11}$$

with $\omega = \phi - \Phi$, in Cartesian (x, y, z) or cylindrical (ρ, ϕ, z) coordinates. The vector field $\boldsymbol{B}$ then becomes

$$\begin{aligned}\boldsymbol{B} &= -2[\nabla(\cos\theta)] \times [\nabla\Phi] = -2[\nabla(\cos\theta)] \times [\nabla(\phi-\omega)] \\ &= -2[\nabla(\cos\theta)] \times \left[\frac{\hat{\boldsymbol{\phi}}}{\rho} - \nabla\omega\right].\end{aligned} \tag{11.12}$$

In general, it may be difficult to solve the differential equation (11.2) for the vector potential $\boldsymbol{A}$. However, this calculation simplifies considerably if the director field has azimuthal symmetry.[2] With azimuthal symmetry, the angles $\theta(\rho, z)$ and $\omega(\rho, z) = \phi - \Phi$ depend only on ρ and z, not on ϕ. In that case, one specific solution for $\boldsymbol{A}$ (one choice of gauge) is

$$\boldsymbol{A} = \frac{2(1-\cos\theta)\hat{\boldsymbol{\phi}}}{\rho} + 2(1+\cos\theta)\nabla\omega. \tag{11.13}$$

The integral (11.3) for the Hopf index then becomes

$$\begin{aligned}Q &= -\frac{1}{(8\pi)^2}\int d\boldsymbol{r}(\boldsymbol{B}\cdot\boldsymbol{A}) = -\frac{1}{(8\pi)^2}\int \rho d\rho d\phi dz \left(\frac{8}{\rho}\right)\hat{\boldsymbol{\phi}}\cdot[\nabla(\cos\theta)\times\nabla\omega] \\ &= \frac{1}{4\pi}\int d\rho dz \sin\theta\left[(\partial_z\theta)(\partial_\rho\omega) - (\partial_\rho\theta)(\partial_z\omega)\right].\end{aligned} \tag{11.14}$$

For a specific example, consider the director field of Eqs. (11.7)–(11.9), based on the Hopf fibration. In cylindrical coordinates, it is parameterized by

$$\theta = 2\sin^{-1}\left(\frac{4\rho}{4+\rho^2+z^2}\right), \quad \omega = \tan^{-1}\left(\frac{4-\rho^2-z^2}{4z}\right). \tag{11.15}$$

An explicit calculation then gives the Hopf index

(continued)

$$Q = \frac{1}{4\pi} \int d\rho dz \frac{256\rho}{(4 + \rho^2 + z^2)^3} = 1. \qquad (11.16)$$

The Hopf index is also $Q = 1$ for the director field compressed into the ball.

[2] J. Gladikowski and M. Hellmund, "Static Solitons with Nonzero Hopf Number," *Phys. Rev. D* **56**, 5194 (1997); see also https://hopfion.com/hopf.html.

11.2 Hopfions as Topological Solitons

In Chaps. 1 and 2, when we developed the concept of winding number as a 1D topological charge, we used this concept to investigate topological defects and solitons. Likewise, in Chap. 10, when we developed the concept of a 2D topological charge, we used that concept to investigate hedgehogs as topological defects, and skyrmions as topological solitons. At this point, now that we have the concept of the Hopf index as a 3D topological charge, we should consider whether it leads to any further types of defects or solitons. In this section, let us begin with solitons.

In most cases, the variation of the director field $\hat{\boldsymbol{n}}(\boldsymbol{r})$ is spread out over the entire size of a system. However, under certain circumstances discussed below, the variation can be concentrated in a small, local region, while the rest of the system has a uniform background orientation. If the localized variation has a Hopf index $Q \neq 0$, it is called a *hopfion*.

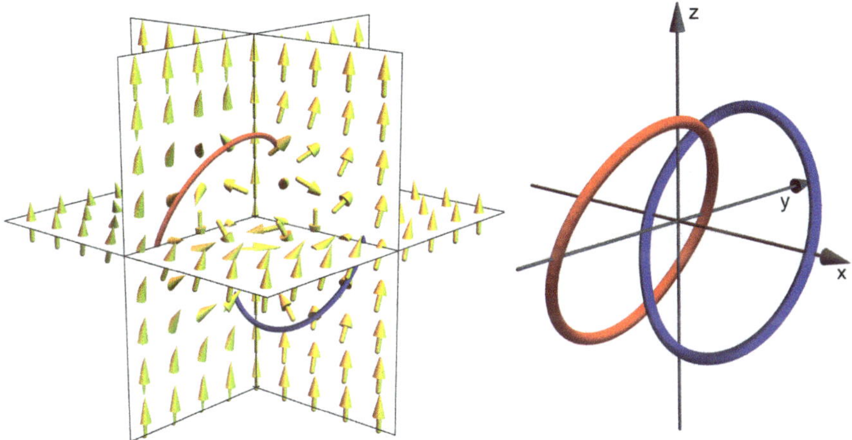

Fig. 11.3 Example of a hopfion, with a localized distortion of the director field $\hat{\boldsymbol{n}}(\boldsymbol{r})$ in an infinite system, with Hopf index $Q = 1$. The red and blue curves represent preimages where $\hat{\boldsymbol{n}}(\boldsymbol{r}) = \hat{\boldsymbol{n}}_1 = (\hat{\boldsymbol{z}} + \hat{\boldsymbol{x}})/\sqrt{2}$ or $\hat{\boldsymbol{n}}_2 = (\hat{\boldsymbol{z}} + \hat{\boldsymbol{y}})/\sqrt{2}$, respectively. The right column provides a full view of the linked preimages in position space; it is *not* a view of director space

One example of a hopfion is shown in Fig. 11.3. In this example, the director field has a complex structure, similar to the Hopf fibration, with linked preimages. The director variations are concentrated in a small region near the origin. Outside that small region, the director goes exponentially toward the uniform background $\hat{\boldsymbol{n}} = \hat{z}$.

Note that a hopfion has a limited size in all three dimensions, x, y, and z. In that sense, a hopfion is effectively a point-like object, which is zero-dimensional. This dimensionality is consistent with Eq. (2.1): For a hopfion in 3D, we have $d_{\text{space}} = 3$ and $d_{\text{measure}} = 3$, and hence $d_{\text{hopfion}} = 0$.

Under what circumstances can the variation of $\hat{\boldsymbol{n}}(\boldsymbol{r})$ be concentrated in a small region to form a hopfion? The requirements are the same as the requirements for a skyrmion, as discussed in Sect. 10.1.3. According to the Derrick-Hobart theorem, this concentrated variation is unstable if the free energy has the usual elastic form of Eq. (10.6). However, it can be stable if the free energy has two additional features: (1) some term that gives orientational alignment of $\hat{\boldsymbol{n}}(\boldsymbol{r})$ in the background orientation, and (2) some term that gives a favorable variation of $\hat{\boldsymbol{n}}(\boldsymbol{r})$, normally chirality giving a favorable twist.

Hopfions have actually been observed in experiments, especially in series of experiments on chiral liquid crystals by Ivan Smalyukh and collaborators. Some of these experiments use liquid crystals doped with colloidal magnetic particles, which have polar order, and other experiments use undoped liquid crystals, which have nematic order. Remarkably, the experiments can image the full 3D configuration of $\hat{\boldsymbol{n}}(\boldsymbol{r})$, and hence visualize the preimages. These experimental preimages are indeed linked, as in the predicted structure of hopfions. The experiments also find knotted structures embedded in a helical background, which have been called *heliknotons*. In some cases, the topological structures are isolated. In other cases, they occur in large numbers and form a 3D crystal. For further theoretical and experimental information on these structures, we refer readers to papers from that research group.[3]

11.3 Topological Defects Based on Hopf Index?

In the previous section, we discussed solitons (hopfions) based on the 3D topological charge (Hopf index). Can there also be defects based on the Hopf index?

For the 1D topological charge (winding number), defects are defined by constructing a 1D measuring path *around* a region in the 2D plane or in 3D space. Likewise, for the 2D topological charge, defects (hedgehogs) are defined by constructing a 2D measuring surface *around* a region in 3D space. For the 3D topological charge (Hopf index), we would need to define defects by constructing a

[3] For early work on hopfions, see B. G. Chen, P. J. Ackerman, G. P. Alexander, R. D. Kamien, and I. I. Smalyukh, "Generating the Hopf Fibration Experimentally in Nematic Liquid Crystals," *Phys. Rev. Lett.* **110**, 237801 (2013). For a 3D crystal, see J. S. B. Tai and I. I. Smalyukh, "Three-Dimensional Crystals of Adaptive Knots," *Science* **365**, 1449 (2019). For a review, see J.-S. Wu and I. I. Smalyukh, "Hopfions, Heliknotons, Skyrmions, Torons and Both Abelian and Nonabelian Vortices in Chiral Liquid Crystals," *Liq. Cryst. Rev.* **10**, 34 (2022).

3D measuring space *around* a region. However, that construction is impossible. A 3D measuring space does not go around any region in 3D space, because it is all of 3D space.

We can also see the same point from Eq. (1.10) for the dimensionality of defects. If $d_{\text{space}} = 3$ and $d_{\text{measure}} = 3$, then d_{defect} would need to be -1, which is impossible.

Some readers might ask: Can we think of time as the fourth dimension, so that a 3D measuring surface can enclose a defect in 4D space-time? Yes, it is possible to think that way. For example, suppose that a material has uniform orientational order, with Hopf index $Q = 0$, for all time $t < 0$. At $t = 0$, a small region of the sample near $\boldsymbol{r} = 0$ is heated into the disordered phase for a short interval of time. After it cools, for $t > 0$, it forms a hopfion with $Q = 1$, still surrounded by uniform orientational order. In that situation, the space-time event at $\boldsymbol{r} = 0$ and $t = 0$ can be regarded as a topological defect. It is a localized event between $Q = 0$ and $Q = 1$, just as the hedgehog in Fig. 10.3 is a localized point between $N = 0$ and $N = 1$. This event in 4D space-time can be surrounded by a curved 3D measuring surface in space-time, which goes before, around, and after the event, and that curved 3D surface has Hopf index $Q = 1$.

We should emphasize that this space-time perspective is *not* the normal way of thinking about defects in condensed-matter physics. Normally, we think of defects as physical objects in space, not as momentary events in space-time. Hence, in the normal perspective, there are no topological defects based on the Hopf index. Rather, there are only hopfions as topological solitons.

Phases with Regular Arrays of Defects or Solitons

12

In *most* of this book, we have considered systems in which the ground state has *uniform* orientational order. The free energy function provides some penalty for any variations in the orientational order. As a result, any topological defects or solitons cost free energy, compared with the uniform state.

There are some important exceptions to this general rule. In these cases, certain variations in orientational order might actually reduce the free energy, relative to uniform orientational order. Hence, certain types of topological defects or solitons might be stable, i.e. lower in free energy than the uniform state. If so, the ground state of the system could have a regular array of defects or solitons.

In previous chapters, we have already seen a few examples of stable defects or solitons. In Sect. 3.5, we considered a 1D system with a combination of chirality and an applied magnetic field, and found that this combination can give stable sine-Gordon solitons. In Sects. 10.1.3 and 10.2.3, we saw that a similar combination of chirality with imposed anisotropy can give stable skyrmions. In Sect. 11.2, we saw that it can give stable hopfions.

In this chapter, we provide a more general view of stable defects or solitons. First, we show that these stable features arise from the general phenomenon of *frustration*, and discuss some simple examples of frustration. Then, we use the concept of frustration to interpret specific physical examples of phases with regular arrays of defects or solitons.

12.1 Concept of Frustration

Many physical effects come from the competition between two different mechanisms. This competition can lead to very different results, depending on whether a physical system has one degree of freedom or many interacting degrees of freedom.

If there is only one degree of freedom, then the competition leads to a simple compromise between the states favored by the two mechanisms. For an example

J. V. Selinger, *Introduction to Topological Defects and Solitons*, Lecture Notes in Physics 1032, https://doi.org/10.1007/978-3-031-70200-6_12

from undergraduate physics, suppose we have a mass hanging from a spring, suspended from the ceiling. The only degree of freedom is the height of the mass. Gravity provides a potential energy of $U_{\text{grav}} = mgz$, which pulls the mass toward the floor. The spring provides a potential energy of $U_{\text{spring}} = \frac{1}{2}k(z_{\text{ceiling}} - z)^2$, which pulls the mass toward the ceiling. Clearly, these two types of potential energy cannot both be minimized at the same time. Rather, the competition between them gives the compromise height of $z = z_{\text{ceiling}} - mg/k$.

By contrast, if there are many interacting degrees of freedom, then the competition might not lead to a simple compromise. Instead, it might lead to an interesting, nonuniform state, which is much more complex than either of the ideal states favored by the two different physical mechanisms. The term *frustration* refers to this type of competition. In the following sections, we will discuss some examples.

12.1.1 Chirality vs. Magnetic Field

For a first example of frustration, let us return to the case of a chiral material in an applied magnetic field, as discussed in Sect. 3.5 and illustrated in Fig. 3.5. As a reminder, the free energy of Eq. (3.26) is

$$F = \int_{-\infty}^{\infty} dz \left[\frac{1}{2} K \left(\frac{d\theta}{dz} \right)^2 + \mu B (1 - \cos\theta) - D \frac{d\theta}{dz} \right]. \tag{12.1}$$

Here, the first term is the standard elastic free energy that penalizes variations of $\theta(z)$, and the second term is the magnetic field coupling that favors alignment of $\hat{\boldsymbol{n}}(z) = (\cos\theta(z), \sin\theta(z), 0)$ along the $+x$-axis. The third term favors a twist of the orientational order. It is called a Dzyaloshinskii-Moriya interaction in the magnetic literature, or a chiral term in the liquid-crystal literature. Note that the second and third terms compete with each other. The orientational order cannot both be aligned along the $+x$-axis and also have a twist. That conflict is the source of frustration in this system.

In Sect. 3.5, we showed that there is a threshold value of the chiral coefficient, $D_{\text{crit}} = 4\sqrt{K\mu B}/\pi$, at which the free energies of the uniform state and the single-soliton state are equal. For $D > D_{\text{crit}}$, the ground state has one or more solitons. Likewise, for $D < -D_{\text{crit}}$, the ground state has one or more antisolitons. If there are multiple solitons or multiple antisolitons, they repel each other and form a regular array.

For further insight into this model, let us consider what happens if $|D| < D_{\text{crit}}$. In that case, the ground state of the *infinite* system is uniform, with $\theta(z) = 0$ everywhere, with no solitons. The free energy of this ground state is $F = 0$. We should emphasize that this uniform ground state is frustrated. It has a natural tendency to twist, but it cannot actually twist anywhere.

How can we tell that the ground state is frustrated? One answer is: Look at a *finite* system with *free* boundaries, and see what happens at the boundaries. In general, boundaries are less constrained than the bulk, so they can reveal structure that is suppressed in the bulk.[1]

For that reason, let us consider a finite system, defined for $-L/2 \le z \le L/2$, with free boundaries at $z = \pm L/2$. From variational calculus, we can derive the Euler-Lagrange equation and boundary conditions

$$\frac{d^2\theta}{dz^2} = \frac{1}{\xi^2}\sin\theta, \text{ with } \xi = \sqrt{\frac{K}{\mu B}}, \tag{12.2a}$$

$$\frac{d\theta}{dz} = \frac{D}{K} \text{ at } z = \pm L/2. \tag{12.2b}$$

To solve these equations, we assume that $\theta(z) \ll 1$ everywhere; we will see that this assumption is self-consistent. The Euler-Lagrange equation can then be linearized as

$$\frac{d^2\theta}{dz^2} = \frac{1}{\xi^2}\theta. \tag{12.3}$$

The solution is

$$\theta(z) = \frac{D}{\sqrt{K\mu B}} \frac{\sinh(z/\xi)}{\cosh(L/(2\xi))}, \tag{12.4}$$

which satisfies $\theta(z) \ll 1$ provided that $D \ll \sqrt{K\mu B}$. This solution is illustrated in Fig. 12.1a. In the bulk, the orientational order is well-aligned with the magnetic field. However, in *boundary layers* of width ξ, the orientational order is able to twist.

By putting the solution for $\theta(z)$ back into Eq. (12.1), we obtain the local free energy density as a function of z, which is plotted in Fig. 12.1b. This result shows that boundary layers make favorable, negative contributions to the total free energy. By comparison, the free energy density goes to zero away from the boundaries. The bulk is frustrated, as it cannot achieve the negative free energy associated with the boundary twist. By integrating over z, we find that the total free energy of the finite system is

$$F = -\frac{D^2\xi}{K}\tanh\frac{L}{2\xi}, \tag{12.5}$$

compared with $F = 0$ for the infinite system. Figure 12.1c shows a plot of the average free energy per length. We can see that it *increases* as a function of system size. Hence, the free energy favors *smaller* systems, with greater surface-to-volume

[1] We thank Eran Sharon for helpful discussions about this point.

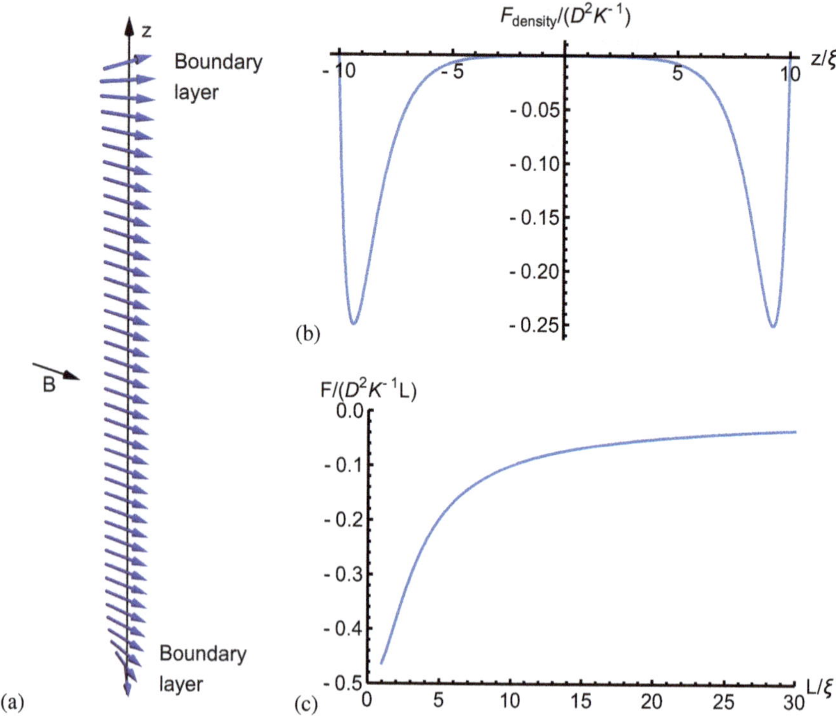

Fig. 12.1 (**a**) Competition between chirality and applied magnetic field in a *finite* system. The system has no twist in the bulk, but it has twist in boundary layers of width ξ at the free surfaces. This figure should be compared with Fig. 3.5 for an infinite system. (**b**) Local free energy density (scaled by parameters D^2K^{-1}) as a function of position z (scaled by ξ). This free energy density is negative in the boundary layers, and zero in the interior. (**c**) Total free energy of the finite system (scaled by D^2K^{-1} and by system size L) as a function of L (scaled by ξ). Note that F/L increases as L increases

ratio. This dependence of free energy on system size is called *superextensive*, and it is a common characteristic of frustrated systems.[2]

This argument gives an intuitive explanation for why solitons are stable at larger values of D: In this system, boundaries are energetically favorable, and solitons are effectively internal boundaries on the orientational order. They break the system into approximately uniform domains, separated by walls. In general, we should expect to see stable solitons or defects in systems where the free energy favors breaking up frustrated domains into smaller regions.

[2] See the discussion of superextensive free energy in S. Meiri and E. Efrati, "Cumulative Geometric Frustration in Physical Assemblies," *Phys. Rev. E* **104**, 054601 (2021).

12.1.2 Packing of Pentagons vs. Euclidean Geometry

In the previous example, frustration is caused by a competition between two physical effects, chirality and field alignment. By comparison, in other situations, frustration might be caused by a single physical effect that is incompatible with the geometry of space. In those cases, it is called *geometric frustration.*

For an example of geometric frustration, let us consider how to tile the 2D plane with regular polygons. If the polygons are equilateral triangles, or squares, or regular hexagons, then the tiling works perfectly. The polygons fit together with their neighbors, and they do not leave any gaps. However, if the polygons are regular pentagons, then there is a serious problem, as illustrated in Fig. 12.2a. We can fit three pentagons around a single point, but they leave a narrow gap, which is too small for a fourth pentagon. Hence, regular pentagons cannot pack together to cover the 2D plane.

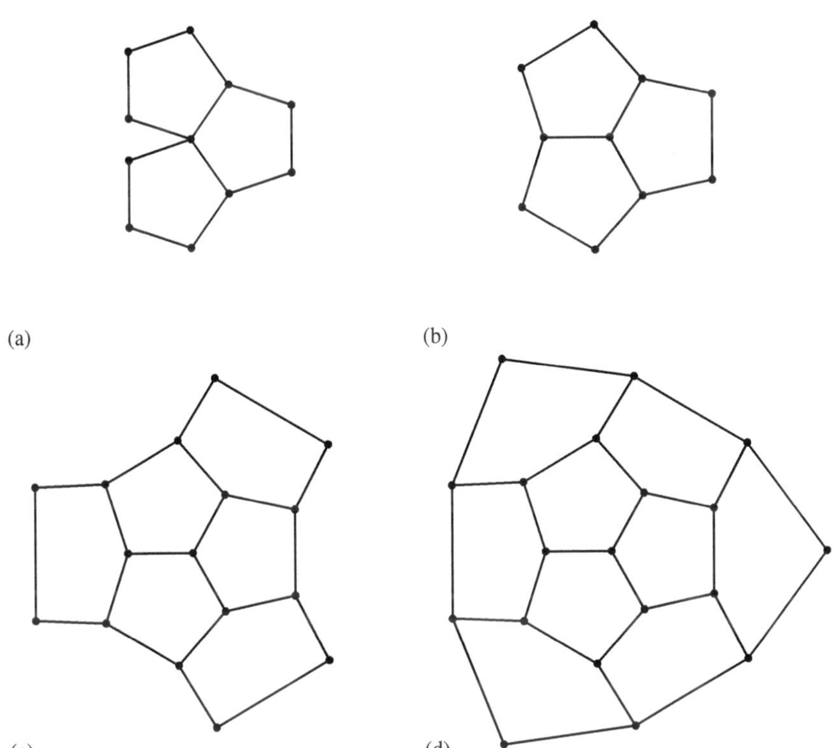

Fig. 12.2 (**a**) Cluster of three regular pentagons, showing that they leave a gap. (**b–d**) Clusters of deformed pentagons. As the clusters grow larger, the necessary amount of distortion increases. These structures are conceptual illustrations; they are not derived from any simulation or energy minimization

What would happen if we try to push the pentagons together? Imagine that the pentagons are made of rubber, so that they can have some elastic deformation, but this deformation costs energy. If we push three pentagons together, they might form the cluster shown in Fig. 12.2b. In this cluster, they are somewhat distorted away from their perfect regular shapes, but not too severely. If we add three more pentagons, we might get the structure in Fig. 12.2c, in which the new pentagons are more distorted. If we add another three pentagons, the cluster might look like Fig. 12.2d. The distortions become even greater as the structure grows larger.

As we go from Fig. 12.2b and c, and further to Fig. 12.2d, we keep adding pentagons with higher distortions and hence higher energies. For that reason, the average energy per area keeps going up as the structure grows larger. This trend is the same superextensive behavior of the energy that we saw in the previous section. It shows that the system of pentagons has geometric frustration, in that the packing of pentagons is not compatible with the Euclidean geometry of the plane.[3] A small number of pentagons might be able to form a cluster, without too much deformation energy. However, this assembly is not possible for a large number of pentagons. The elastic distortion energy prevents the growth of a small cluster into a large cluster.

In a system of many pentagons, we would expect to form many small clusters separated by *something*. They might just be separated by empty space, leading to many unconnected clusters. Alternatively, they might be separated by some type of solitons or defects. We are not aware of any theoretical or experimental results for this specific problem of packing pentagons. However, the general phenomenon of particle assembly with geometric frustration has been widely studied. It provides a common mechanism for the *self-limited growth*, leading to assemblies with a specific finite size.[4]

12.1.3 Favored Splay (or Bend) vs. Euclidean Geometry

The concept of geometric frustration is not limited to packing problems; it also applies to orientational order. To demonstrate this phenomenon, let us consider a 2D polar phase with some favored splay. For example, this favored splay could occur in a system of wedge-shaped particles. If one end of the particles takes up more space than the other end, then a state with nonzero splay would have a lower free energy than a uniform state.

To model this system, we can use the 2D director field $\hat{\boldsymbol{n}}(\boldsymbol{r}) = (\cos\theta(\boldsymbol{r}), \sin\theta(\boldsymbol{r}))$ with the free energy

[3] By comparison, the packing of pentagons *is* compatible with the non-Euclidean geometry of a sphere, with the right ratio of the sphere radius to the pentagon edge length. Twelve pentagons can pack perfectly on a sphere to form a dodecahedron.

[4] For a review of this general phenomenon, see G. M. Grason, "Perspective: Geometrically Frustrated Assemblies," *J. Chem. Phys.* **145**, 110901 (2016).

$$F = \int dr \left[\frac{1}{2} K |\nabla \hat{\boldsymbol{n}}|^2 - \lambda \nabla \cdot \hat{\boldsymbol{n}} \right] \tag{12.6a}$$

$$= \int dr \left[\frac{1}{2} K (\nabla \cdot \hat{\boldsymbol{n}})^2 + \frac{1}{2} K (\nabla \times \hat{\boldsymbol{n}})^2 - \lambda \nabla \cdot \hat{\boldsymbol{n}} \right] \tag{12.6b}$$

$$= \int dr \left[\frac{1}{2} K \left(\nabla \cdot \hat{\boldsymbol{n}} - \frac{\lambda}{K} \right)^2 + \frac{1}{2} K (\nabla \times \hat{\boldsymbol{n}})^2 - \frac{\lambda^2}{2K} \right] \tag{12.6c}$$

$$= \int dr \left[\frac{1}{2} K |\nabla \theta|^2 - \lambda \hat{z} \times \hat{\boldsymbol{n}} \cdot \nabla \theta \right]. \tag{12.6d}$$

From Eq. (12.6c), we can see that the optimal local state occurs when

$$\nabla \cdot \hat{\boldsymbol{n}} = \frac{\lambda}{K}, \quad \nabla \times \hat{\boldsymbol{n}} = 0. \tag{12.7}$$

However, there is no director field $\hat{\boldsymbol{n}}(\boldsymbol{r})$ that satisfies Eqs. (12.7) everywhere in the plane. These equations are incompatible with the Euclidean geometry of the plane.[5] Hence, this system is another case of geometric frustration.

To minimize the free energy under geometric frustration, consider a strip with $-L/2 \leq x \leq L/2$, infinite in y. Figure 12.3a shows the state with the lowest free energy that can actually be achieved. Along the midline $x = 0$, the director field satisfies Eqs. (12.7) exactly. Away from the midline, the deviation from these equations gradually increases. This state can be approximately described as $\theta(x) \approx \pi/2 - \pi x/L$, and hence the average free energy per area can be estimated as

$$\frac{F}{A} \approx \frac{\pi^2 K}{2L^2} - \frac{2\lambda}{L}. \tag{12.8}$$

For width $L > \pi^2 K/(2\lambda)$, the average free energy per area increases as L increases. Once again, this trend is a superextensive dependence on system size. It shows that the free energy favors an optimum strip width.

If the system is large, then it should form finite regions of splay separated by *something:*

- They might just be separated by empty space, leading to many unconnected strips of splay.

[5] The incompatibility was proven mathematically by I. Niv and E. Efrati, "Geometric Frustration and Compatibility Conditions for Two-Dimensional Director Fields," *Soft Matter* **14**, 424 (2018). This article also shows that the equations *are* compatible with the non-Euclidean geometry of a hyperbolic (saddle-like) surface. Likewise, if the favored state has zero splay and zero bend, it is incompatible with any non-Euclidean geometry that has nonzero Gaussian curvature, as discussed in Sect. 6.3.

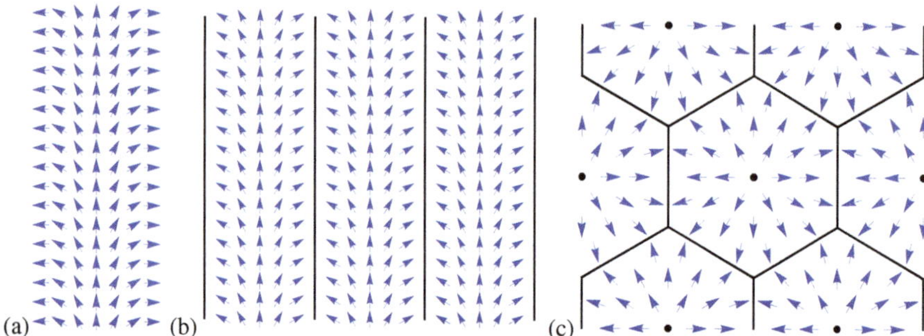

Fig. 12.3 Possible configurations for 2D orientational order with favored splay. (**a**) Single strip, finite in x, infinite in y. (**b**) 1D periodic structure with regions of splay separated by orientational domain walls. (**c**) 2D periodic structure with hexagonal domains of splay separated by orientational domain walls. In structure (**c**), there is a defect of $q = +1$ at the center of each domain, and a defect of $q = -1/2$ in the nematic order at each vertex where three domains meet. In a model with favored bend, the configurations would look exactly the same, but with all arrows rotated by 90°

- They might be separated by orientational domain walls, as shown in Fig. 12.3b. In this structure, the system has splay across a finite width of x, and then a wall where the orientational order jumps back. This sequence of splay and walls repeats periodically. The walls are similar to the walls shown in Fig. 7.8a. As mentioned in Sect. 7.2, there can be different microscopic models for what happens inside a wall. For example, the polar order might deviate from the nematic order, or its magnitude might pass through zero.
- As an alternative possibility, the system might form a honeycomb lattice of splay domains, illustrated in Fig. 12.3c. Here, the hexagonal domains are separated by orientational domain walls. At the center of each domain, there is a defect of winding number $q = +1$. At each vertex where three domains come together, we can interpret the configuration as a defect of $q = -1/2$, if we think about the nematic rather than the polar order. Hence, this structure actually provides a stable lattice of topological defects.

In this section, we have discussed a 2D model with favored *splay* of $\hat{\boldsymbol{n}}(\boldsymbol{r})$. Exactly the same considerations also apply to a 2D model with favored *bend* of $\hat{\boldsymbol{n}}(\boldsymbol{r})$. To go between these models, we only need to rotate $\hat{\boldsymbol{n}}(\boldsymbol{r})$ by 90°, i.e. rotate all of the arrows in Fig. 12.3 by 90°. This rotation exchanges the roles of 2D splay $(\nabla \cdot \hat{\boldsymbol{n}})$ and 2D bend $(\nabla \times \hat{\boldsymbol{n}})$.

One physical realization of the favored-bend model is in thin films of chiral smectic-C liquid crystals. In a smectic-C phase, the molecules form layers in the (x, y) plane, and they are tilted with respect to the layer normal $\hat{z}$. The director $\hat{\boldsymbol{n}}(\boldsymbol{r})$ has both horizontal and vertical components. Researchers then define $\boldsymbol{c}(\boldsymbol{r})$ as the projection of $\hat{\boldsymbol{n}}(\boldsymbol{r})$ into the (x, y) plane. If the molecules are chiral, the 3D vector field $\hat{\boldsymbol{n}}(\boldsymbol{r})$ has a favored twist. This favored twist of $\hat{\boldsymbol{n}}(\boldsymbol{r})$ corresponds to a favored bend of the 2D vector field $\boldsymbol{c}(\boldsymbol{r})$. Hence, the theory of this section applies to $\boldsymbol{c}(\boldsymbol{r})$.

Both the favored-splay and favored-bend models have been studied over many years of liquid-crystal research.[6]

12.2 Examples of Phases with Defects or Solitons

In this section, we present several physical examples of phases with regular arrays of defects or solitons, and discuss the origin of these phases in terms of frustration.

12.2.1 Lattices of Skyrmions or Half-Skyrmions (Merons)

In Chap. 10, we already showed that chiral magnets and liquid crystals can form lattices of skyrmions or half-skyrmions (merons). Here, we review those results using the concept of frustration.

First, consider thin films of chiral magnets, as discussed in Sect. 10.1.3. In these films, the free energy has two special features, in addition to the standard elastic free energy. One feature is an orientational alignment of $\hat{\boldsymbol{n}}(\boldsymbol{r})$ in a background orientation. This alignment might be caused by an applied magnetic field or by lattice anisotropy. The second feature is a favored twist of $\hat{\boldsymbol{n}}(\boldsymbol{r})$, caused by the Dzyaloshinskii-Moriya interaction. These two features of the free energy compete with each other. The orientational order cannot both be aligned in the background orientation and also have a twist. For that reason, this 2D system is frustrated, just like the 1D system in Sects. 3.5 and 12.1.1.

The system may "relieve the frustration" in several possible ways; i.e. its ground state may have several possible structures:

1. It might form a state that is uniformly aligned in the background orientation everywhere in the interior. This structure would be the ground state for *weak* Dzyaloshinskii-Moriya interaction. If the system is finite with free boundaries, it would have twist of $\hat{\boldsymbol{n}}$ on the edges.
2. It might form a state with a 1D lattice of sine-Gordon solitons in $\hat{\boldsymbol{n}}$ as a function of a single coordinate x, uniformly extended in the orthogonal y-direction. This structure would look like Fig. 3.5b–d, but lying on its side, so that $\hat{\boldsymbol{n}}$ depends on x and has a background orientation in the z-direction.

[6] To our knowledge, the first study of favored splay was R. B. Meyer and P. S. Pershan, "Surface Polarity Induced Domains in Liquid Crystals," *Solid State Commun.* **13**, 989 (1973). The theory of stripes in chiral smectic-C films was developed by S. A. Langer and J. P. Sethna, "Textures in a Chiral Smectic Liquid-Crystal Film," *Phys. Rev. A* **34**, 5035 (1986). The possibility of a hexagonal lattice was considered by G. A. Hinshaw, R. G. Petschek, and R. A. Pelcovits, "Modulated Phases in Thin Ferroelectric Liquid-Crystal Films," *Phys. Rev. Lett.* **60**, 1864 (1988). We discussed several versions of this theory in the review article R. D. Kamien and J. V. Selinger, "Order and Frustration in Chiral Liquid Crystals," *J. Phys. Condens. Matter* **13**, R1 (2001).

3. It might form a state with a 2D lattice of skyrmions in $\hat{\boldsymbol{n}}$ as a function of x and y. This structure is shown in Fig. 10.8a. As we have discussed, skyrmions are 2D versions of solitons. Like sine-Gordon solitons, skyrmions provide a compromise between the background alignment and the favored twist.
4. It might form a state with a 2D lattice of half-skyrmions (merons) in $\hat{\boldsymbol{n}}$ as a function of x and y, illustrated in Fig. 10.8b. In this structure, each half-skyrmion must be accompanied by a topological defect of winding number $q = -1$. Hence, it can only be the ground state if the free energy cost of a defect is not too high.

Possibilities 2 and 3 can be regarded as lattices of solitons, while possibility 4 can be regarded as a lattice of defects. The free energy balance between these possibilities depends on the strength of the background alignment, the Dzyaloshinskii-Moriya interaction, the free energy cost of defects, and perhaps also other parameters. Hence, there may a complex phase diagram for the range of stability of each possible structure.

Second, consider thin films of chiral liquid crystals, as discussed in Sect. 10.2.3. Once again, the free energy involves both orientational alignment and favored chiral twist of $\hat{\boldsymbol{n}}(\boldsymbol{r})$. In this case, the orientational alignment is usually provided by surface anchoring on the top and bottom of a liquid-crystal cell. If the top and bottom substrates have homeotropic (perpendicular) anchoring conditions, they require $\hat{\boldsymbol{n}} = \pm\hat{z}$. It is impossible for the director field to satisfy these anchoring conditions and also have the favored twist everywhere inside the cell. Hence, this system experiences geometric frustration, arising from the geometry of the cell and its surfaces.

In response to this geometric frustration, the system may form several different possible ground states, similar to the previous list:

1. Uniformly aligned state, perhaps with twist of $\hat{\boldsymbol{n}}$ on the edges if there are free boundaries.
2. 1D lattice of sine-Gordon solitons.
3. 2D lattice of skyrmions, illustrated in Fig. 10.15a.
4. 2D lattice of half-skyrmions (merons), with each half-skyrmion accompanied by two topological defects of winding number $q = -1/2$, as shown in Fig. 10.15b.

Again, possibilities 2 and 3 are lattices of solitons, and possibility 4 is a lattice of defects. The phase diagram for these possible structures may be quite complex.

12.2.2 Blue Phases

Next, let us consider a chiral liquid crystal in an infinite 3D geometry, with no surface anchoring or magnetic field. Naively, this system might not seem to be frustrated, because chirality does not compete with any other physical mechanism.

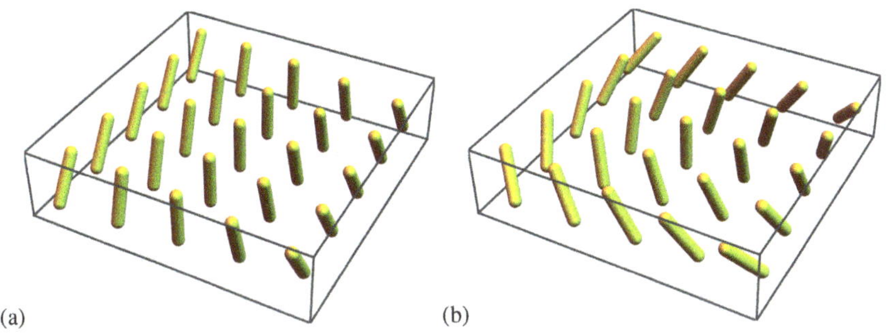

Fig. 12.4 Comparison between twisted states of a director field. (**a**) Single twist. (**b**) Double twist

However, it actually experiences geometric frustration arising from the conflict between chiral twist and 3D Euclidean space.

To see this frustration, we must consider the distinction between single and double twist. In single twist, $\hat{\boldsymbol{n}}$ varies as a function of a single coordinate x, as shown in Fig. 12.4a. It can be represented in Cartesian coordinates (x, y, z) as

$$\hat{\boldsymbol{n}}_{\text{single}} = \hat{\boldsymbol{z}} \cos qx - \hat{\boldsymbol{y}} \sin qx. \tag{12.9}$$

In double twist, $\hat{\boldsymbol{n}}$ varies as a function of two coordinates x and y, as shown in Fig. 12.4b. It can be represented in cylindrical coordinates (ρ, ϕ, z) as

$$\hat{\boldsymbol{n}}_{\text{double}} = \hat{\boldsymbol{z}} \cos q\rho - \hat{\boldsymbol{\phi}} \sin q\rho \text{ for } q\rho \ll 1. \tag{12.10}$$

We put both of these expressions into the free energy density of a chiral liquid crystal,

$$F^{\text{density}} = \frac{1}{2} K |\nabla \hat{\boldsymbol{n}}|^2 + D \hat{\boldsymbol{n}} \cdot \nabla \times \hat{\boldsymbol{n}}, \tag{12.11}$$

and obtain

$$F^{\text{density}}_{\text{single}} = \frac{1}{2} K q^2 - Dq, \qquad F^{\text{density}}_{\text{double}} = K q^2 - 2Dq \text{ for } q\rho \ll 1. \tag{12.12}$$

For the moment, let us consider only the local region $q\rho \ll 1$, and minimize the free energy density there. This local minimization gives $q = D/K$ for both single and double twist. At that minimum, the free energy density becomes

$$F^{\text{density}}_{\text{single}} = -\frac{D^2}{2K}, \qquad F^{\text{density}}_{\text{double}} = -\frac{D^2}{K} \text{ for } q\rho \ll 1. \tag{12.13}$$

Hence, the optimal *local* state of a chiral liquid crystal is double twist, not single twist.[7]

Single twist can fill up 3D Euclidean space perfectly. The equations above for single twist are valid at all positions. However, double twist cannot fill up 3D Euclidean space. The equations above for double twist are valid only for $\rho \ll 1/q = K/D$ in cylindrical coordinates. Beyond that radius, the director field does not have the favored twist. Hence, a chiral liquid crystal experiences geometric frustration; it has an optimal local structure that is not compatible with 3D Euclidean space.[8] Indeed, as an explicit test for geometric frustration, we have recently simulated a chiral liquid crystal with free boundaries.[9] In those simulations, the interior has single twist, and a boundary layer has double twist, which is analogous to the boundary-layer effect discussed in Sect. 12.1.1. This result confirms that the system is frustrated.

Under this geometric frustration, how can a chiral liquid crystal in an infinite 3D geometry minimize its total free energy, integrated over all space? The most common answer is to form a cholesteric phase. A cholesteric phase has the director field of Eq. (12.9), with single twist. It is a compromise between the double twist favored by chirality and the constraints of Euclidean geometry. However, it is not the only possible answer.

As an alternative, a chiral liquid crystal can form a blue phase. A blue phase is a complex structure consisting of cylindrical regions of double twist, often called *double-twist tubes* or *double-twist cylinders*, separated by disclinations. It is interesting to note that the double-twist tubes do not need to be separated by domain walls. They can be just be separated by disclination lines in 3D, with the director field continuous around the disclinations.

One might guess that the structure of a blue phase would look like the lattice of half-skyrmions (merons), shown in Fig. 10.15b, uniformly extended into the third dimension. Actually, the structure is more complex than that. The double-twist tubes are not all parallel to each other. Rather, they form a periodic lattice with cubic symmetry, and the disclinations run diagonally between these tubes. There are two versions of this cubic structure: blue phase I with body-centered cubic symmetry, shown in Fig. 12.5a, and blue phase II with simple cubic symmetry,

[7] Here, for simplicity, we have used a free energy with only a single elastic constant. The same comparison can also be done using the full Oseen-Frank free energy with four elastic constants. In that case, double twist is preferred over single twist because of the K_{24} term. For this analysis, see J. V. Selinger, "Interpretation of Saddle-Splay and the Oseen-Frank Free Energy in Liquid Crystals," *Liq. Cryst. Rev.* **6**, 129 (2018).

[8] By comparison, double twist *is* compatible with the non-Euclidean geometry of S^3, which is a 3D sphere embedded in 4D space. For the theory in non-Euclidean geometry, see J. P. Sethna, D. C. Wright, and N. D. Mermin, "Relieving Cholesteric Frustration: The Blue Phase in a Curved Space," *Phys. Rev. Lett.* **51**, 467 (1983); J. F. Sadoc, R. Mosseri, and J. V. Selinger, "Liquid Crystal Director Fields in Three-Dimensional Non-Euclidean Geometries," *New J. Phys.* **22**, 093036 (2020).

[9] C. Long and J. V. Selinger, "Explicit Demonstration of Geometric Frustration in Chiral Liquid Crystals," *Soft Matter* **19**, 519 (2023).

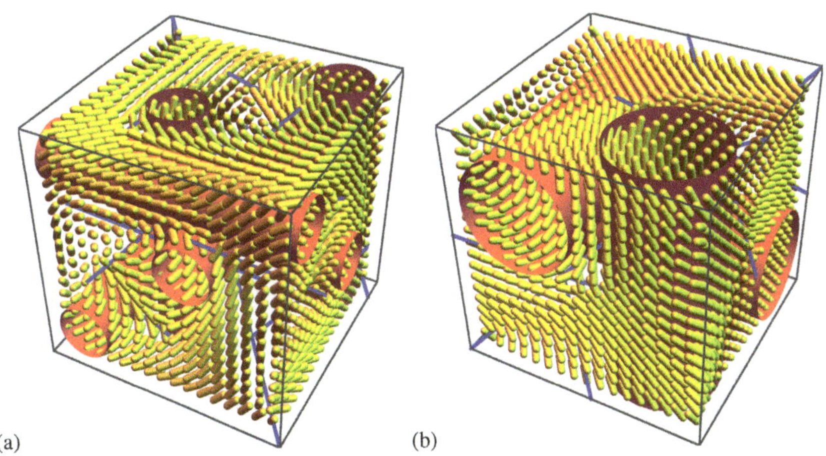

Fig. 12.5 Structure of blue phases. (**a**) Blue phase I, with body-centered cubic symmetry. (**b**) Blue phase II, with simple cubic symmetry. In these visualizations, the small yellow cylinders represent the director field. The large red cylinders are just guides to the eye, indicating the double-twist tubes; they are *not* physical objects. The diagonal blue lines are the disclinations, which are physical objects

shown in Fig. 12.5b. There is also a blue phase III, not shown, which is a disordered fluid of double-twist tubes and disclinations.[10]

There have been many detailed theoretical studies of the phase diagram for cholesteric and blue phases. We believe that the essential physics can be explained by the following argument: Compared with the cholesteric phase, the advantage of forming a blue phase is that it has double twist rather than single twist. This benefit is an average free energy density of order Kq^2. However, the disadvantage of forming a blue phase is that it has disclination lines. Within a cubic unit cell of volume $a^3 \sim q^{-3}$, the total length of disclination lines is of order $a \sim q^{-1}$. Hence, the average free energy density of the disclination lines is $(F_{\text{disclination}}a)/a^3 \sim F_{\text{disclination}}q^2$, where $F_{\text{disclination}}$ is the disclination line tension (free energy per length). Hence, a blue phase is favored if $F_{\text{disclination}}$ is sufficiently small compared with the elastic constant K.[11]

To summarize the results of this section, chirality leads to a favored double-twist deformation, which leads to geometric frustration. If the disclination line tension is

[10] J. Pišljar, S. Ghosh, S. Turlapati, N. V. S. Rao, M. Škarabot, A. Mertelj, A. Petelin, A. Nych, M. Marinčič, A. Pusovnik, M. Ravnik, and I. Muševič, "Blue Phase III: Topological Fluid of Skyrmions," *Phys. Rev. X* **12**, 011003 (2022). The double-twist tubes are sometimes regarded as fractional skyrmions, because they have the same geometric structure as twisted skyrmions but the director only covers a fraction of the unit sphere. They do not have the same topology as full skyrmions.

[11] In a model with unequal Frank elastic constants, a blue phase is favored if $F_{\text{disclination}}$ is sufficiently small compared with K_{24}.

high, the liquid crystal forms a cholesteric phase as a compromise structure, which can fill up space without defects. If the disclination line tension is low, the liquid crystal forms a blue phase, which is a periodic lattice of double-twist tubes and disclinations.

By the same argument, other types of molecular order lead to other favored director deformations, which also lead to geometric frustration. In a recent review article, we have discussed the complex modulated structures that can form under those conditions.[12] Many of these structures have been observed experimentally.

12.2.3 Superconductors and Twist-Grain-Boundary Phases

Outside of liquid crystals and magnetism, the Abrikosov vortex lattice in superconductors is the best-known example of a phase with a periodic array of topological defects. The basic theory of the vortex lattice is quite consistent with the general framework of this chapter.

One important feature of superconductivity is the *Meissner effect:* Superconductors exclude magnetic fields. Suppose that a sample is initially in the high-temperature normal (non-superconducting) state. When a small magnetic field is applied, magnetic flux passes through the sample. If the sample is cooled below the critical temperature T_c into the low-temperature superconducting state, the magnetic flux is expelled from the interior. Most of the magnetic flux goes outside the sample, and some passes through a narrow surface region of thickness λ, known as the London penetration depth.

The Meissner effect shows that the superconducting state is incompatible with a magnetic field. If a magnetic field is applied to a superconductor, the material is *frustrated.* The optimal local state of the material has superconducting order, and the optimal local state also has magnetic flux. These two requirements are inconsistent with each other. The competition between these requirements is similar to the competition between chirality and a magnetic field in Sect. 12.1.1, and the London penetration depth is similar to the boundary layer of twist in Fig. 12.1.

Now suppose that the strength of the magnetic field is increased. The resulting behavior depends on the type of superconductor. In a *type I* superconductor, the superconducting state continues up to a critical magnetic field H_c. Beyond H_c, the material has a phase transition back to the normal state, and magnetic flux passes through the sample. By contrast, in a *type II* superconductor, the behavior is more complex. At a critical field H_{c1}, the material forms a hexagonal lattice of vortices. These vortices are extended tubes, aligned with the magnetic field, and each vortex carries a quantum of magnetic flux. Hence, the magnetic flux is not completely

[12] J. V. Selinger, "Director Deformations, Geometric Frustration, and Modulated Phases in Liquid Crystals," *Annu. Rev. Condens. Matter Phys.* **13**, 49 (2022). As a clarification: In that article, we used the word "defects" to mean any non-ideal local structures. In the terminology of this book, we would call those structures either defects or solitons. We regret this change of terminology.

expelled from the sample; some passes through the vortices. This vortex lattice continues up to a higher critical field H_{c2}, when the material has a phase transition back to the normal state.

Vortices in type II superconductors are analogous to topological defects in the xy model, which we have discussed in this book. The superconducting order parameter is a complex number, with a magnitude and a phase. If we go around a loop about any vortex, the phase rotates through 2π. Each vortex has a characteristic core radius, the superconducting coherence length ξ, inside which the magnitude of superconducting order goes to zero. This behavior is similar to the magnitude of xy order, modeled in Chap. 5. Hence, each vortex can be regarded as a small region of the normal state, surrounded by the superconducting state. In a thin film of superconductor, the vortices are defect points in 2D. In a bulk superconductor, they are lines in 3D. The lattice of vortices is then a periodic array of defects, which is stable because of the frustration between magnetic field and superconducting order.

In 1972, Pierre-Gilles de Gennes pointed out a mathematical analogy between the theory of superconductors and the theory of smectic liquid crystals.[13] In smectic liquid crystals, the molecules form layers but remain fluid within each layer. Because of that structure, these phases are effectively crystalline in one dimension. As crystals, they can have dislocations, with the Burgers vector normal to the layers. De Gennes showed that the smectic order parameter is a complex number, like the superconducting order parameter. Furthermore, the free energy of smectic liquid crystals couples gradients in smectic order with nematic director fluctuations, just as the Ginzburg-Landau theory of superconductors couples gradients of the order parameter with magnetic vector potential fluctuations. Based on that analogy, he suggested that some smectic liquid crystals should be in type II, with a structure like the Abrikosov vortex lattice, but with a periodic array of dislocation lines.

Several years later, Scot Renn and Tom Lubensky developed a detailed theory inspired by de Gennes' suggestion.[14] In their theory, chiral twist is analogous to magnetic field. Twist is incompatible with smectic order, just as magnetic field is incompatible with superconducting order. Hence, a chiral smectic liquid crystal is frustrated; it cannot achieve both twist and smectic order. In response to this frustration, twist might be concentrated in *screw dislocations*, which we discussed in Sect. 9.2.2.

One difference between superconducting vortices and screw dislocations is that screw dislocations cannot all be parallel to each other. Because of this difference, a lattice of screw dislocations must have the structure shown in Fig. 12.6. In this structure, the material has a series of smectic domains, with strong smectic order and no twist. Between these domains, there are grain boundaries, with twist and reduced smectic order. Each grain boundary is composed of a series of screw dislocations.

---

[13] P. G. de Gennes, "An Analogy between Superconductors and Smectics A," *Solid State Commun.* **10**, 753 (1972).

[14] S. R. Renn and T. C. Lubensky, "Abrikosov Dislocation Lattice in a Model of the Cholesteric–to-Smectic-A Transition," *Phys. Rev. A* **38**, 2132 (1988).

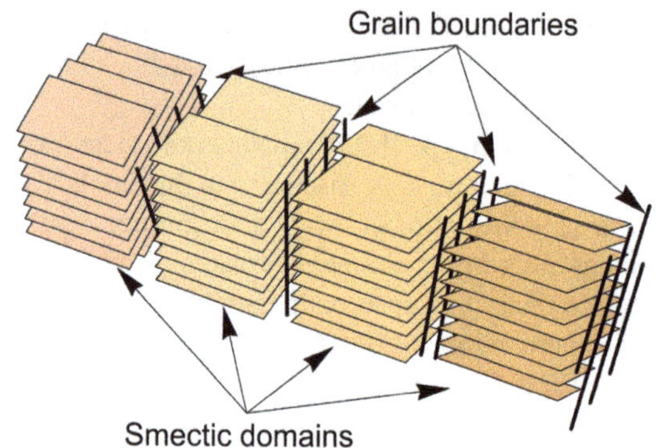

Fig. 12.6 Structure of a twist-grain-boundary (TGB) phase, with alternating smectic domains and grain boundaries. Each grain boundary consists of a series of screw dislocations, indicated by the thick black lines

Renn and Lubensky called this structure the *twist grain boundary (TGB) phase*. They developed a phase diagram with analogues of H_{c1} and H_{c2}, showing the transitions among the cholesteric, TGB, and uniform smectic phases.

The TGB phase has been observed experimentally using a range of techniques, including x-ray diffraction and freeze-fracture microscopy. More complex variations have also been studied theoretically and experimentally, including versions based on tilted smectic-C as well as untilted smectic-A order. The TGB structure provides a notable example of how frustration can lead to phases with regular arrays of defects.

Index

J. V. Selinger, *Introduction to Topological Defects and Solitons*, Lecture Notes in Physics 1032, https://doi.org/10.1007/978-3-031-70200-6

GPSR Compliance
The European Union's (EU) General Product Safety Regulation (GPSR) is a set of rules that requires consumer products to be safe and our obligations to ensure this.

If you have any concerns about our products, you can contact us on

ProductSafety@springernature.com

In case Publisher is established outside the EU, the EU authorized representative is:

Springer Nature Customer Service Center GmbH
Europaplatz 3
69115 Heidelberg, Germany

www.ingramcontent.com/pod-product-compliance
Ingram Content Group UK Ltd.
Pitfield, Milton Keynes, MK11 3LW, UK
UKHW021840270726
14058UKWH00002B/259
9783031701993